COMPOSTING IN THE CLASSROOM

SCIENTIFIC INQUIRY FOR HIGH SCHOOL STUDENTS

NANCY M. TRAUTMANN
Center for the Environment, Cornell University

MARIANNE E. KRASNY
Department of Natural Resources, Cornell University

KENDALL/HUNT PUBLISHING COMPANY
4050 Westmark Drive   Dubuque, Iowa 52002

ABOUT THE AUTHORS:

Ms. Nancy Trautmann is an Extension Associate with the Cornell Center for the Environment. **Dr. Marianne Krasny** is an Associate Professor in the Cornell Department of Natural Resources and leader of the Cornell Program in Environmental Sciences for Educators and Youth. Together they work with high school teachers, Cornell scientists, and community educators to create opportunities for students to conduct original research in the environmental sciences.

WITH CONTRIBUTIONS BY HIGH SCHOOL TEACHERS:

- Patrick Cushing
- Stephanie Hyson
- Alpa Khandar
- Richard Northrup
- Elaina Olynciw
- Barbara Poseluzny
- Timothy Sandstrom

AND RESEARCH SCIENTISTS:

- Erin McDonnell
- Tom L. Richard

PRODUCTION:

Cover and Page Design:	Jane MacDonald
Illustration:	Lucy Gagliardo
	Jane MacDonald
Copy Editing:	Laura Glenn
Photography:	PhotoSynthesis Productions, Inc.
	Nancy Trautmann
	Elaina Olynciw
	Erin McDonnell

Background cover photo by KPT Power Photos.

FUNDING:

 National Science Foundation

 Cornell Waste Management Institute

 Cornell Center for the Environment

Copyright © 1998 by Nancy M. Trautmann and Marianne E. Krasny

Library of Congress Catalog Card Number: 97-74994

ISBN 0-7872-4433-3

All rights reserved. No part of this publication may be reproduced, stored in a retrieval system, or transmitted, in any form or by any means, electronic, mechanical, photocopying, recording, or otherwise, without the prior written permission of the copyright owner.

Printed in the United States of America

10 9 8 7 6 5 4 3

ACKNOWLEDGMENTS

For four summers, high school science teachers conducted composting research at Cornell University and developed plans for using composting as a topic for scientific inquiry by their students. *Composting in the Classroom* evolved from the experiences of these teachers and the Cornell faculty and staff with whom they worked.

We would like to thank the following Cornell scientists for serving as mentors to teachers in their composting research projects: Eric Nelson, John Peverly, Joe Regenstein, Larry Walker, Erin McDonnell, and Cheryl Craft. Without their dedicated assistance, this work would not have been possible.

Tom Richard and Erin McDonnell provided inspiration and invaluable technical advice during all phases of preparation of this book. In addition, we are grateful to the following individuals for technical review of portions of the manuscript: Patrick Bohlen, Jean Bonhotal, Veet Deha, Bennett Kottler, Fred Michel, and Eric Nelson. Insights into school composting and student research were provided by Timothy Conner, Mark Johnson, Maddalena Polletta, and Richard Realmuto. Mark Johnson and his Ithaca High School students graciously allowed us to photograph their classroom composting experiments. Thank you to Ellen Harrison for her invaluable support, both moral and financial, and to Veronique Organisoff and Denise Weilmeier for their assistance with logistics.

Finally, we wish to thank our families for cheerfully enduring all manner of composting experiments, in our kitchens, gardens, and yards, as we tested the procedures described in this book.

Funding was provided by the National Science Foundation, the Cornell Waste Management Institute, and the Cornell Center for the Environment.

CONTENTS

PREFACE VII

INTRODUCTION IX
   Composting Systems IX
   Composting Health and Safety X
   Wastes or Resources? XI

1 THE SCIENCE OF COMPOSTING 1
   Thermophilic Composting 2
   Compost Chemistry 5
      Chemical Requirements for Thermophilic Composting 6
   Compost Physics 9
      Mechanisms of Heat Loss 9
      Aeration 10
      Moisture 11
      Particle Size 11
      Size of Compost System 12
   Compost Biology 13
      Microorganisms 14
      Invertebrates 17
      Earthworms 22

2 COMPOSTING BIOREACTORS AND BINS 27
   Two-Can Bioreactors 28
   Soda Bottle Bioreactors 31
   Worm Bins 35
   Outdoor Composting 41
      Holding Units 41
      Turning Units 42
      Enclosed Bins 42

3 GETTING THE RIGHT MIX 43
   Choosing the Ingredients: General Rules of Thumb 43
      Moisture 43
      Carbon-to-Nitrogen Ratio 44
      Other Considerations 46
   Calculations for Thermophilic Composting 47
      Moisture 47
      Carbon-to-Nitrogen Ratio 48

4 MONITORING THE COMPOSTING PROCESS 51
   Temperature 52
   Moisture 52
   Odor 53
   pH 53
   Microorganisms 55
      Observing Compost Microorganisms 55
      Culturing Bacteria 57

　　　　Culturing Actinomycetes 60
　　　　Culturing Fungi 62
　　　　Measuring Microbial Activity 64
　　Invertebrates. 66
　　　　Pick and Sort 67
　　　　Berlese Funnel 68
　　　　Wet Extraction 69

5　COMPOST PROPERTIES 71
　　Compost Stability. 73
　　　　Jar Test 73
　　　　Self-Heating Test. 74
　　　　Respiration Test 75
　　Compost Quality 79
　　　　Phytotoxicity Bioassay 79
　　Effects on Soil Properties 83
　　　　Porosity 83
　　　　Water Holding Capacity 85
　　　　Organic Matter Content 87
　　　　Buffering Capacity 89

6　COMPOST AND PLANT GROWTH EXPERIMENTS 91
　　Plant Growth Experiments 93

7　COMPOSTING RESEARCH 97
　　Exploration and Controlled Experiments 98
　　Narrowing Down a Research Question 99
　　Example Research Projects 100
　　Interpreting Results 102
　　Final Words 103

GLOSSARY 105

FOR MORE INFORMATION 111

INDEX 115

PREFACE

The science education reform movement in the United States and other countries is challenging high school science teachers to rethink the way in which they guide student learning. New standards call for students to learn science in a manner similar to the way science is practiced, including conducting original inquiry and research.

The goal of *Composting in the Classroom* is to provide high school science teachers with the background needed to engage students in research focusing on composting. There are a number of reasons why composting research lends itself well to the classroom setting. First, composting of yard wastes and food scraps presents a partial solution to the solid waste crisis; thus, composting research addresses practical problems of concern to students and their communities. Second, nearly all the equipment and materials are inexpensive and readily available. With proper maintenance to prevent odor and insect problems, composting systems can be set up in the classroom, as well as outdoors in the school yard. Finally, many experiments can be conducted within two weeks or less, although long-term composting research projects are also possible.

Perhaps more important is the pedagogical rationale for conducting classroom composting research. To start, students can generate an unlimited number of questions about the composting process, and they can design their own experiments to answer these questions. Because there is much about the science of composting that remains unknown, well-designed student research projects can contribute to the existing body of knowledge. In order to conduct these investigations, students must draw on their understanding of scientific concepts from a variety of disciplines. Conducting composting research requires an understanding of biological, chemical, and physical processes, such as uptake of carbon and nitrogen by microorganisms, diffusion of oxygen through air and water, and effect of moisture on heat production and transfer. Designing compost systems also provides opportunities to bring technology into the science classroom, and to discuss the interface between science and society, and science and the students' personal lives.

Composting in the Classroom begins with an overview of composting science (Chapter 1). Chapters 2 and 3 provide instructions on how to build and add the right mix of ingredients to compost systems. Chapter 4 outlines how to monitor the composting process, and Chapter 5 describes how to measure the attributes of the finished compost. Once students have made and tested their compost, they might want to use it in plant growth experiments (Chapter 6). Several important points about conducting research are included in Chapter 6 as well as in Chapter 7. This last chapter is a discussion of how to help students design meaningful research projects focusing on composting.

The information included in the following pages reflects what the scientific community knows about composting as of 1997. But scientific knowledge is constantly changing, as people make new observations and conduct research to validate or invalidate these observations. Teachers and students using this manual are challenged to use their own observations to question the information that is presented, and to design and conduct research to test their observations. At first, that research may be exploratory in nature, designed to explore in a relatively unsystematic fashion any number of variables that may affect the composting process. Based on the results of their exploratory research, students may identify one or two variables that they would like to test in a controlled experiment. Or they may wish to use their knowledge about composting to create an optimal composting system as a technological design project.

*Throughout the manual, we have presented "**Research Possibilities**" in italics. These are ideas that students could develop into exploratory and controlled experimental research and technological design projects. They are meant as suggestions only; students are encouraged to come up with their own original research and design questions.*

Seeing other students conducting composting research may help inspire your students to engage in similar investigations. You may wish to obtain a copy of *It's Gotten Rotten*, a video that provides an introduction to the science of composting and shows students performing composting experiments (see list of resources on page 111).

School composting is carried out for two distinct purposes: for disposal of leaves, grass clippings, or food scraps, and as a focus for scientific exploration by students. The intent of this manual is to provide teachers and students with the background necessary for using composting as a focus for scientific inquiry. For information on how to set up composting systems designed to handle organic wastes for an entire school, refer to *Composting...Because a Rind Is a Terrible Thing to Waste* (Bonhotal and Rollo, 1996) or *Scraps to Soil: A How-to Guide for School Composting* (Witten, 1995).

Many of the activities in *Composting in the Classroom* were developed by a group of high school teachers who spent several summers conducting research at Cornell University, and then engaged their classroom and science club students in similar research projects. This manual is dedicated to these and other innovative teachers who have accepted the challenge to help students learn about science by engaging them in open-ended inquiry.

Nancy Trautmann
Marianne Krasny

INTRODUCTION

The decomposition of organic materials is as ancient as nature itself. Without decomposition, nutrients would be locked into existing organic matter, unavailable for growth of new organisms. Composting, which refers to the controlled decomposition of organic materials, has been used by farmers and gardeners since prehistoric times to recycle wastes and make them available for plant growth. However, since World War II, as U.S. farms have become larger and more mechanized, the use of compost and other traditional means of enhancing soil productivity has declined. In recent years, concern about reducing solid waste and producing food in an environmentally sound manner has led to a renewed interest in composting.

Organic material, including food scraps and yard trimmings, constitutes from 20% to 40% of the total waste stream in the U.S. Composting presents an opportunity to keep these materials out of landfills and incinerators. To realize this opportunity, however, composters must find answers to many questions about the composting process. How can we balance air flow and moisture to create the ideal conditions for decomposition of any particular mixture of organic materials? What mixtures of organic materials represent carbon-to-nitrogen ratios that promote the growth of decomposer microorganisms? How can we tell if compost is "finished" and ready to be used in greenhouses or gardens? Because there are so many unanswered questions, students can be part of the process of obtaining scientific information about composting. The results of the students' investigations can be used to design better composting systems in their own home, school, or community.

COMPOSTING SYSTEMS

Composting can be carried out in many different ways, ranging from using small, indoor worm bins that process a few pounds of food per week to huge commercial or industrial operations that process many tons of organic matter in long, outdoor piles called windrows. Many systems rely solely on microorganisms, largely bacteria and fungi, to do the work of decomposition. In these systems, successful composting depends on creating conditions that are favorable to the growth and activity of microbial communities. Vermicomposting, or composting with worms, relies on the action of microorganisms and any of several species of worms that are known to thrive in environments high in organic matter. Many outdoor composting systems are invaded by soil invertebrates, which, similar to worms in vermicomposting, play a role in decomposition together with bacteria and fungi.

Because a by-product of microbial activity is heat, compost systems will become hot if they supply adequate aeration, moisture, and nutrients for rapid microbial growth and are sufficiently large or insulated to retain the heat produced. For example, the temperature in the middle of an outdoor compost pile may rise to 55°C within the first few days of mixing the organic materials, even when surrounding air temperatures are below freezing. Hence, the name for this type of composting is *thermophilic*, or "heat-loving."

Composting also occurs at lower temperatures over longer periods of time. For example, a pile of autumn leaves will gradually decompose over the course of a year or two without ever getting hot, either because the nitrogen or moisture level is insufficient or because the pile is too small to be adequately self-insulating. In this case, decomposition is carried out by *mesophilic* (moderate temperature) microorganisms and invertebrates. Similarly, vermicomposting does not produce the high temperatures found in thermophilic composting, and in fact, such temperatures would be lethal to the worms. Although the microbes in vermicompost generate heat, it is rapidly dissipated in the relatively shallow compost bins or troughs needed to provide optimal conditions for surface-dwelling worms.

In addition to outdoor piles, composting can be conducted in containers or vessels called bioreactors. Large-scale bioreactors are used in commercial and industrial composting, whereas small-scale bioreactors are useful for conducting research within the space available in labs or small outdoor research areas. Two types of bioreactors are described in this manual: the two-can system composed of nested, plastic garbage cans, and a smaller system built from soda bottles. Bioreactors similar to these are used by engineers to model the processes of air flow, heat transfer, and enzymatic degradation that occur during composting.

COMPOSTING HEALTH AND SAFETY

There are two potential hazards in working with compost. The first relates to the type of organic materials that are being composted, some of which may contain disease-causing organisms, or pathogens. Plate scrapings, meat and dairy products, and pet wastes should not be used in classroom composting because of their potential to spread disease or attract pests. To avoid student exposure to pathogens, it is best to use only yard and garden trimmings and pre-consumer vegetable and fruit scraps. Many school cafeterias assist classrooms conducting compost investigations by saving vegetable peels and other materials generated in preparing salads and vegetable and fruit dishes. Even when using only the recommended organic materials, students should take normal sanitary measures when handling compost (wash hands when finished, avoid touching eyes or mouths).

The second potential composting hazard relates to allergic reactions to airborne spores. This is not a widespread problem. Studies of workers in large-scale compost facilities and of individuals who live near such facilities have failed to find increased incidence of allergic reactions or any other health problems. However, just as individuals vary in their resistance to disease, a few individuals may be particularly sensitive to some of the fungi in compost. One of these fungal species, *Aspergillus fumigatus*, can infect the respiratory system of a sensitive person who is heavily exposed to it. (Although this fungus is ubiquitous, it occurs in relatively high concentrations in compost.) Conditions that may predispose individuals to infection or an allergic response include: a weakened immune system, punctured eardrum, allergies, asthma, and use of some medications such as antibiotics and adrenal cortical hormones. People with these conditions should avoid mixing and prolonged contact with compost, and they should consider wearing dust masks.

Flies and odors in compost may be an annoyance, but generally they do not pose a health problem. With proper maintenance of the compost pile, these annoyances can be avoided (see Chapters 2–4).

WASTES OR RESOURCES?

Are the organic materials that are used to produce compost really "wastes," or are they valuable "resources" that should be conserved and recycled? Some environmentalists object to the use of the term *waste*, feeling that it implies that the food scraps and yard trimmings many people throw in the garbage are useless. Rather, these individuals prefer to use terms that are neutral (e.g., scraps, trimmings, material) or even terms that imply these materials are of great value (e.g., resources). We have tried whenever practical to avoid the use of the term wastes, and to promote the idea of recycling these materials, although the word waste does appear occasionally throughout the text. You may want to use the issue of appropriate terminology—wastes or resources—as a starting point for a classroom discussion on composting.

1
THE SCIENCE OF COMPOSTING

If moist food scraps are placed in a container and left to sit for a week or two, the end product is likely to be a smelly "slop" that attracts flies. Given the proper conditions, these same food scraps can be composted to produce a material that looks and smells like rich soil and can be used to enhance soil texture and productivity. So, what are these conditions that promote composting?

The physical and chemical conditions that should be maintained in a compost heap seem logical if you think about what compost really is—a big pile of food for billions of minute organisms. These microorganisms have certain chemical requirements, primarily carbon for energy, nitrogen to build proteins, and oxygen for respiration. Interacting with these are physical requirements, such as aeration to maintain optimal oxygen levels while not depleting the moisture necessary for microbial growth.

This chapter begins with an overview of the chemical, physical, and biological changes that occur during thermophilic composting, thereby providing an example of how several sciences can be integrated in the study of composting. This overview is followed by more detailed, separate sections on compost chemistry, physics, and biology. Much of the information in the chemistry and physics sections focuses on thermophilic composting. The biology section includes a discussion of the microbes that are present in all types of composting, and of the diversity of invertebrates that live in outdoor and some worm composting systems.

The information presented here reflects the current body of scientific knowledge regarding composting. Much remains unknown, leaving a variety of intriguing questions for future research. Some of these questions are identified as **Research Possibilities**, sprinkled in italics throughout the text. We hope they will provide the inspiration for students to ask a much wider range of questions, and to design their own original research.

THERMOPHILIC COMPOSTING

Many composting systems are based on providing the optimal conditions for thermophilic composting because its high temperatures promote rapid decomposition and kill weed seeds and disease-causing organisms. These high temperatures are a by-product of the intense microbial activity that occurs in thermophilic composting. Thermophilic composting can be divided into three phases, based on the temperature of the pile: (1) a mesophilic, or moderate-temperature phase (up to 40°C), which typically lasts for a couple of days; (2) a thermophilic, or high-temperature phase (over 40°C), which can last from a few days to several months depending on the size of the system and the composition of the ingredients; and (3) a several-month mesophilic curing or maturation phase. Periodic temperature measurements can be used to chart the progress of thermophilic composting, producing a "temperature profile" showing these three phases (Figure 1–1).

Figure 1–1. The Three Phases of Thermophilic Composting.

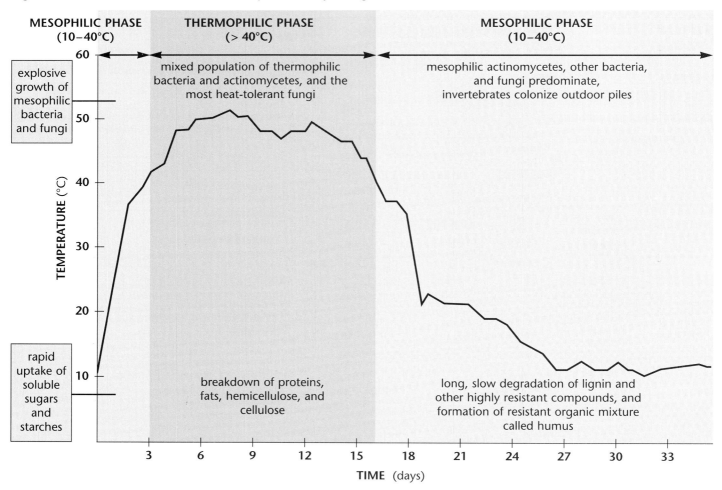

Different communities of microorganisms predominate during the various temperature phases. Initial decomposition is carried out by mesophilic microorganisms, those that thrive at moderate temperatures. These microbes rapidly break down the soluble, readily degradable compounds, and the heat they produce causes the compost temperature to rapidly rise. Once temperatures exceed 40°C, the mesophilic microorganisms become less competitive and are replaced by thermophilic, or heat-loving microbes. During the thermophilic stage, high temperatures accelerate the breakdown of proteins, fats, and complex carbohydrates like cellulose and hemicellulose, the major structural molecules in plants. As the supply of these compounds becomes exhausted, the compost temperature gradually decreases and mesophilic microorganisms once again take over for the final phase of "curing," or maturation of the remaining organic matter. Although the compost temperature is close to ambient during the curing phase, chemical reactions continue to occur that make the remaining organic matter more stable and suitable for plant use.

The smaller systems used for indoor composting are not likely to get as hot as compost in large piles or windrows. A well-designed indoor compost system, ≥10 gallons in volume, will heat up to 40–50°C in the first two or three days. Soda bottle bioreactors, because they are so small, are more likely to peak at temperatures of 40–45°C. At the other end of the scale, commercial or municipal-scale compost systems may reach temperatures above 60°C. Because temperatures above 55°C are lethal to many microorganisms that cause human or plant diseases, this is the target temperature that compost managers use for suppression of pathogens. However, if the compost temperature goes above 60–65°C, the beneficial microbial populations are also killed.

Humans can control the temperatures during composting by mixing or turning the organic materials (Figure 1–2). If the pile or windrow is getting too hot, turning a pile can release heat from the inner core, which temporarily cools it down (points A and B in Figure 1–2). As the food available to thermophilic organisms becomes depleted, their rate of growth slows and the temperature begins to drop. Turning the pile at this point may produce a new temperature peak (points C and D in Figure 1–2). This is because relatively undecomposed organic matter gets mixed into the center of the pile, where temperature and moisture conditions are optimal for rapid decomposition. In addition, mixing loosens up the compost ingredients, which increases the infiltration of oxygen that is needed by aerobic microorganisms. After the thermophilic phase is completed, the compost temperature drops, and it is not restored by turning or mixing (point E).

Figure 1–2. The Effects of Turning on Compost Temperature.
(See p.3 for explanation.)

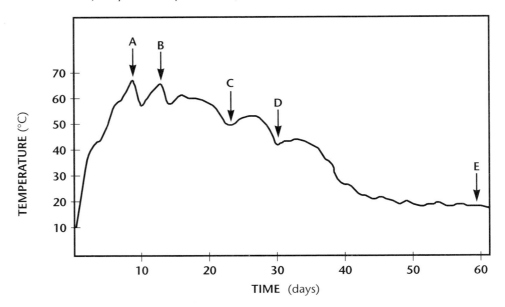

 Research Possibility: *Garden supply stores and catalogs often sell compost "starters," which they claim speed up the composting process. Develop a recipe for a compost starter and design a research project to test its effect on the compost temperature profile. (Hint: You might want to include finished compost or soil as an inoculant, or nitrogen fertilizer or sugar to trigger fast microbial growth.)*

COMPOST CHEMISTRY

Many chemical changes occur during composting, either relatively rapidly in thermophilic systems or more slowly in worm bins or other systems that do not heat up (Figure 1–3). In all of these compost systems, chemical breakdown is triggered by the action of enzymes produced by microorganisms. Bacteria and fungi secrete enzymes that break down complex organic compounds, and then they absorb the simpler compounds into their cells. The enzymes catalyze reactions in which sugars, starches, proteins, and other organic compounds are oxidized, ultimately producing carbon dioxide, water, energy, and compounds resistant to further decomposition. The enzymes are specialized, such as cellulase to break down cellulose, amylase for starches, and protease for proteins. The more complex the original molecule, the more extensive the enzyme system required to break it down. Lignins, large polymers that cement cellulose fibers together in wood, are among the slowest compounds to decompose because their complex structure is highly resistant to enzyme attack.

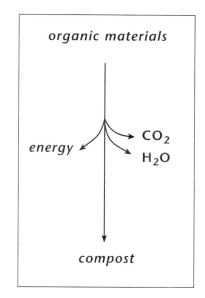

As organic matter decomposes, nutrients such as nitrogen, phosphorus, and potassium are released and recycled in various chemical forms through the microorganisms and invertebrates that make up the compost food web. Proteins decompose into amino acids such as glycine or cysteine. These nitrogen- and sulfur-containing compounds then further decompose, yielding simple inorganic ions such as ammonium (NH_4^+), nitrate (NO_3^-), and sulfate (SO_4^{2-}) that become available for uptake by plants or microorganisms.

Not all compounds get fully broken down into simple ions. Microbes also link some of the chemical breakdown products together into long, intricate chains called polymers. These resist further decomposition and become part of the complex organic mixture called humus, the end product of composting.

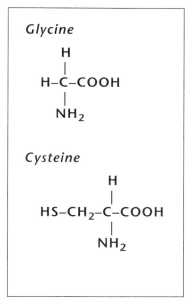

In thermophilic composting, any soluble sugars in the original mixture are almost immediately taken up by bacteria and other microorganisms. The resulting explosive microbial growth causes the temperature to rise. During the thermophilic phase, more complex compounds such as proteins, fats, and cellulose get broken down by heat-tolerant microbes. Eventually, these compounds become depleted, the temperature drops, and the long process of maturation begins. During this final phase, complex polymers continue slowly to break down. Those most resistant to decay become incorporated into humus.

Research Possibility: How well do human nutrition concepts apply to compost microorganisms? For example, will the microbes get a "sugar high," demonstrated by a quick, high temperature peak when fed sugary foods, compared with a longer but lower peak for more complex carbohydrates?

Figure 1–3. Chemical Decomposition during Thermophilic Composting.

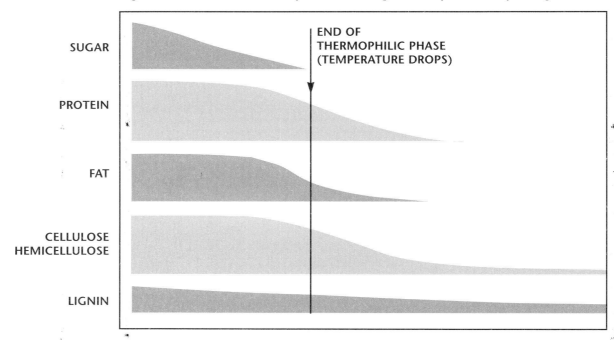

CHEMICAL REQUIREMENTS FOR THERMOPHILIC COMPOSTING

In order for successful thermophilic composting to occur, the proper conditions need to be created for optimal microbial growth. The key factors are the relative amounts of carbon and nitrogen, the balance between oxygen and moisture content, and pH.

CARBON-TO-NITROGEN RATIO

Of the many elements required for microbial decomposition, carbon and nitrogen are the most important and the most commonly limiting. Carbon is both an energy source (note the root in our word for high-energy food: *carbo*hydrate), and the basic building block making up about 50% of the mass of microbial cells. Nitrogen is a crucial component of the proteins, amino acids, enzymes, and DNA necessary for cell growth and function. Bacteria, whose biomass is over 50% protein, need plenty of nitrogen for rapid growth.

The ideal carbon-to-nitrogen (C:N) ratio for composting is generally considered to be around 30:1, or 30 parts carbon for each part nitrogen by weight. Why 30:1? Although the typical microbial cell is made up of carbon and nitrogen in ratios as low as 6:1, additional carbon is needed to provide the energy for metabolism and synthesis of new cells. C:N ratios lower than 30:1 allow rapid microbial growth and speedy decomposition, but excess nitrogen will be lost as ammonia gas, causing undesirable odors as well as loss of the nutrient. C:N ratios higher than 30:1 do not provide sufficient nitrogen for optimal growth of the microbial populations. This causes the compost to remain relatively cool and to degrade slowly, at a rate determined by the availability of nitrogen.

As composting proceeds, the C:N ratio gradually decreases from 30:1 to 10–15:1 for the finished product. This occurs because each time that organic compounds are consumed by microorganisms, two-thirds of the carbon is lost to the atmosphere as CO_2 gas, while most of the nitrogen is recycled into new microorganisms. Although finished compost has a low C:N ratio, this does not result in the odor problems mentioned above because the organic matter is in a stable form, having already undergone extensive decomposition.

Attaining a C:N ratio of roughly 30:1 in the mix of compost ingredients is a useful goal; however, this ratio may need to be adjusted according to the bioavailability of the materials in question. Most of the nitrogen in compostable materials is readily available. Some of the carbon, however, may be bound up in compounds that are highly resistant to biological degradation. Newspaper, for example, decays less readily than other types of paper because it has not been chemically treated to remove lignin. Lignin, a highly resistant compound found in wood, forms sheaths around cellulose fibers, retarding their decomposition. The result is that it takes almost four times as much newsprint as office paper to provide the same amount of bioavailable carbon in composting.[1] Corn stalks and straw are similarly slow to break down because they are made up of a resistant form of cellulose. All of these materials can still be composted, but it is best to mix them with other sources containing more readily biodegradable carbon.

Research Possibility: *Newspaper has a higher lignin content than office paper and therefore should take longer to decompose. Is this true in thermophilic composting? What about in worm bins?*

Particle size can also affect the availability of carbon. Whereas the same amount of carbon is contained in comparable masses of wood chips and sawdust, the larger surface area in the sawdust makes its carbon more readily available for microbial use. A larger volume of wood chips than sawdust would therefore be needed to achieve the same amount of available carbon.

In addition to carbon and nitrogen, adequate phosphorus, sulfur, calcium, and potassium are essential to microbial metabolism, as are trace elements such as magnesium, iron, and copper. Normally, these nutrients are not limiting because the compost ingredients provide sufficient quantities for microbial growth.

OXYGEN

Oxygen is essential for the metabolism and respiration of aerobic microorganisms and for oxidizing the various organic molecules present in the waste material. As microorganisms oxidize organic matter for energy and nutrition, oxygen is used and carbon dioxide is produced. If oxygen supplies are depleted, the composting process will become anaerobic and produce undesirable odors, including the rotten-egg smell of hydrogen sulfide gas. Therefore, compost systems need to be designed to provide adequate air flow using either passive or forced aeration systems.

pH

During the course of composting, the pH generally varies between 5.5 and 8.5 (Figure 1–4). The initial pH depends on the composition of the ingredients. In the early stages of composting, organic acids may accumulate as a by-product of the digestion of organic matter by bacteria and fungi. The resulting drop in pH encourages the growth of fungi, which are active in the decomposition of lignin and cellulose. Usually, the organic acids break down further during the composting process, and the pH rises. This is caused by two processes that occur during the thermophilic phase: decomposition and volatilization of organic acids, and release of ammonia by microbes as they break down proteins and other organic nitrogen sources. Later in the composting process, the pH tends to become neutral as the ammonia is either lost to the atmosphere or incorporated into new microbial growth. Finished compost generally has a pH between 6 and 8.

If the system becomes anaerobic, it will not follow this trend. Instead, acid accumulation may lower the pH to 4.5, severely limiting microbial activity. In such cases, aeration usually is sufficient to return the compost pH to acceptable ranges.

Figure 1–4. Change in pH during Thermophilic Composting.

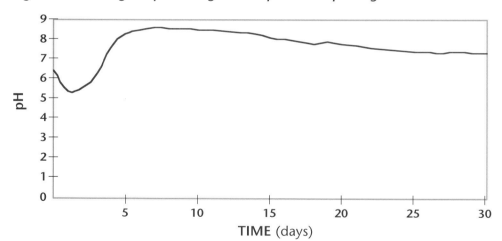

Research Possibility: *Measure the pH of a number of different compost mixes. How does the pH of initial ingredients affect the pH of finished compost?*

Research Possibility: *Some instructions call for adding lime to increase the pH when compost ingredients are mixed. Other instructions caution to avoid this because it causes a loss of nitrogen. How does adding various amounts of lime to the initial ingredients affect the pH of finished compost?*

Research Possibility: *Does the pH of the initial compost ingredients affect the populations of microorganisms during composting?*

COMPOST PHYSICS

Providing the right conditions for thermophilic organisms involves a series of balancing acts between various physical properties of the compost system and its ingredients. The pile or vessel must be large enough to retain heat and moisture, yet small enough to allow good air circulation. The compost must be sufficiently moist to support microbial growth but not so wet that it becomes anaerobic. Also, the particle size of organic materials must be large enough to maintain a porous mix but not so large that decomposition is inhibited.

MECHANISMS OF HEAT LOSS

The temperature at any point during composting depends on how much heat is being produced by microorganisms, balanced by how much is being lost through conduction, convection, and radiation (Figure 1–5). *Conduction* refers to energy that is transferred from atom to atom by direct contact. At the edges of a compost pile, conduction causes heat loss to the surrounding air molecules. The smaller the bioreactor or compost pile, the greater the surface area-to-volume ratio, and therefore, the larger the degree of heat loss to conduction. Insulation helps to reduce this loss in small compost bioreactors.

Figure 1–5. Three Mechanisms of Heat Loss from a Thermophilic Compost Pile.

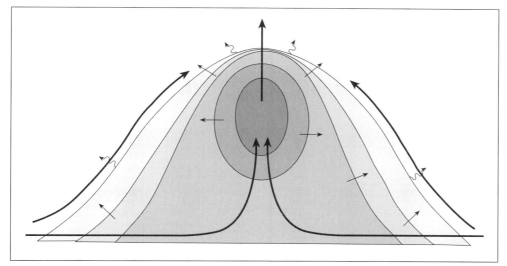
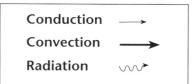

Research Possibility: *Design a thermophilic composting system to provide a heat exchange system for heating water. Compare the effectiveness of several different systems.*

Convection refers to the transfer of heat by movement of a substance such as air or water. When compost gets hot, warm air rises within the system, and the resulting convective currents cause a slow but steady movement of heated air upward through the compost and out the top. In addition to this natural convection, some composting systems use "forced convection" driven by blowers or fans. This forced air, in some cases triggered by thermostats that indicate when the piles are getting too hot, increases the rates of both conductive and convective heat losses. Much of the energy transfer is in the form of latent heat—the energy required to evaporate water. You can sometimes see steamy water vapor rising from hot compost piles or windrows.

The third mechanism for heat loss, *radiation*, refers to electromagnetic waves like those that you feel when standing in the sunlight or near a warm fire. Similarly, the warmth generated in a compost pile radiates out into the cooler surrounding air. However, radiation is a negligible loss of heat from compost because of the relatively small difference in temperature between the outer edges of the compost and the surrounding air.

Research Possibility: *What type of insulation works best for soda bottle bioreactors? Does it help to have a reflective layer? Do different insulative materials or different thicknesses affect the temperature profile?*

Because water has a higher specific heat than most other materials, drier compost mixtures tend to heat up and cool off more quickly than wetter mixtures, providing that adequate moisture levels for microbial growth are maintained. The water acts as a thermal stabilizer, damping out the changes in temperature as microbial activity ebbs and flows.

AERATION

Maintaining the proper balance between moisture and oxygen is one of the keys to successful composting. Because oxygen diffuses thousands of times faster through air than through water, oxygen transfer is impeded if water fills the pores between compost particles. If the thin films of water surrounding individual particles dry out, however, the microorganisms that decompose inorganic matter will become inactive. Therefore, the key to successful composting is to provide enough water to maintain the thin films around compost particles, but not so much that it replaces air in the larger pores.

At the start of the composting process, the oxygen concentration in the pore spaces is about 15–20% (similar to the normal composition of air), and the CO_2 concentration varies from 0.5–5%. As biological activity progresses, the O_2 concentration decreases and CO_2 concentration increases. If the average O_2 concentration in the pile falls below 5%, regions of anaerobic conditions develop. Providing that the anaerobic activity is kept to a minimum, the compost pile acts as a biofilter to trap and degrade the odorous compounds produced as a by-product of anaerobic decomposition. However, should the anaerobic activity increase above a certain threshold, undesirable odors may result.

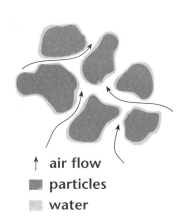

Oxygen concentrations greater than 10% are considered optimal for aerobic composting. Some systems are able to maintain adequate oxygen passively, using air holes or aeration tubes. Others require forced aeration provided by blowers or agitators.

A common misconception in composting is that piles should be turned or mixed every couple of weeks in order to maintain adequate oxygen levels. In fact, calculations show that the oxygen introduced by turning a rapidly degrading compost pile becomes depleted within the first several hours, indicating that diffusion and convection rather than turning are the primary mechanisms that keep the compost aerobic.[2] Mixing does help to keep the pile aerated, however, by loosening it up and increasing the pore spaces through which air flow occurs. Another reason for turning compost is to mix the drier and cooler materials from the edges into the center of the pile, where the more constant heat and moisture promote optimal decomposition.

Research Possibility: When constructing compost bins or piles, some people incorporate perforated pipe, wire mesh, or other systems to increase passive air flow. What is the effect of different methods of aeration on the temperature profile of any one compost system?

Research Possibility: How do various means and schedules for turning a pile affect the temperature profile and the time needed for production of finished compost?

Research Possibility: What is the effect of forced aeration (with an aquarium pump or similar apparatus) on the temperature profile in a soda bottle or a two-can bioreactor?

MOISTURE

An initial moisture content of 50–60% by weight is generally considered optimum for composting because it provides sufficient water to maintain microbial growth but not so much that air flow is blocked. Decomposition by microorganisms occurs most rapidly in the thin films of water surrounding compost particles. When conditions become drier than 35–40%, bacterial activity is inhibited because these films begin to dry up. At the other end of the range, moisture levels above 65% result in slow decomposition, odor production in anaerobic pockets, and nutrient leaching.

PARTICLE SIZE

Most microbial activity occurs on the surface of the organic particles. Therefore, decreasing particle size, through its effect of increasing surface area, will encourage microbial activity and increase the rate of decomposition. Decreasing particle size also increases the availability of carbon and nitrogen. Thus, the carbon in wood shavings or sawdust is more available than the carbon in large wood chips. However, when particles are too small and compact, air circulation through the pile is inhibited. This decreases the oxygen available to microorganisms, and it ultimately decreases the rate of microbial activity and decomposition. "Bulking agents" consisting of large particles such as wood chips, chopped branches, pine cones, or corn cobs are often added to piles to enhance aeration. At the end of the composting process, bulking agents that have not decomposed can be sieved out from the compost and reused.

Research Possibility: Try several bulking agents with different particle sizes. Is there a difference in the temperature profile and length of time it takes to produce compost?

SIZE OF COMPOST SYSTEM

A compost pile must be large enough to prevent rapid dissipation of heat and moisture, yet small enough to allow good air circulation. Conventional wisdom for thermophilic composting is that piles should be at least 1 m³ in size to ensure sufficient heat and moisture retention. Smaller systems such as soda bottle bioreactors require insulation for heat retention. Compost in bioreactors made of nested garbage cans will heat up without insulation if the inner can is at least 10 gallons in size and the surrounding air temperatures are not too low.

Research Possibility: Try mixing the same ingredients in a large outdoor pile, a two-can bioreactor, and a soda bottle bioreactor. Which system reaches the hottest temperatures? Which remains hot the longest? How does this affect the compost produced?

COMPOST BIOLOGY

No matter what system is used for composting, biological organisms play a central role in the decomposition process. The most vital of these are microorganisms, but worms and other invertebrates are also key players in some types of composting.

Food webs provide one way of portraying interactions among organisms. When you think of a food web, you may think of the sun's energy being converted into food by green plants, which are eaten by herbivores, who in turn are eaten by an array of predators. The leaves, feathers, and excrement produced by each of these organisms, as well as the plants and animals themselves when they die, provide the energy source for another type of food web—the decomposition food web. Among leaves and logs on the forest floor, in a steaming pile of hay or manure, or in a compost pile, many of the same organisms are at work. These microorganisms convert organic debris into a source of energy and nutrients for other organisms, as well as serve as prey for higher level microbes, invertebrates, and vertebrates (Figure 1–6).

Figure 1–6. Functional Groups of Organisms in a Compost Food Web.

Tertiary Consumers
organisms that eat secondary consumers
centipedes, predatory mites,
rove beetles, pseudoscorpions

Secondary Consumers
organisms that eat primary consumers
springtails, feather-winged beetles,
and some types of mites,
nematodes, and protozoa

Primary Consumers
organisms that feed on organic residues
actinomycetes and other bacteria, fungi,
snails, slugs, millipedes, sowbugs, some types
of mites, nematodes, and protozoa

Organic Residues
leaves, grass clippings, other plant debris,
food scraps, fecal matter and animal bodies including
those of soil invertebrates

The organic residues forming the base of the compost food web are consumed by fungi, actinomycetes and other bacteria, and invertebrates including millipedes, sowbugs, nematodes, snails, slugs, and earthworms. These primary consumers serve as food for secondary consumers including springtails and some predatory species of nematodes, mites, and beetles. Finally, there are higher-level predators such as centipedes, rove beetles, and pseudoscorpions that prey on each other and the secondary-level compost invertebrates.

The interactions between organic matter, microbes, and invertebrates in compost are sometimes difficult to portray in a food web (Figure 1–7). For example, some invertebrates digest only feces, or organic matter that has already passed through the guts of other organisms. Others feed on fresh organic matter but require microorganisms inhabiting their gut to break it down into a form they can digest. Worms and some other invertebrates derive nutrition by digesting the microorganisms growing on organic detritus, as well as the detritus itself.

Figure 1–7. Feeding Interactions among Organisms in Compost.

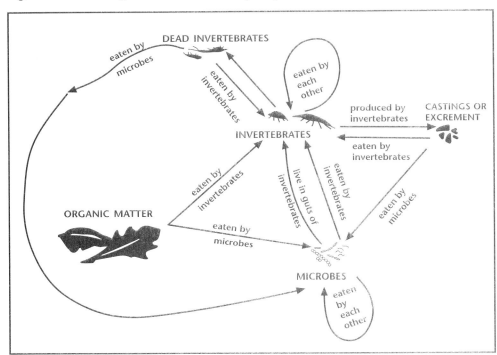

MICROORGANISMS

All types of composting depend on the work of bacteria and fungi. These microbes digest organic matter and convert it into chemical forms that are usable by other microbes, invertebrates, and plants. During thermophilic composting, the populations of various types of microorganisms rise and fall in succession, with each group thriving while environmental conditions and food sources are favorable, then dying off and leaving a new set of conditions that favor another group of organisms. Even in vermicomposting and outdoor composting, microorganisms play an active role, within the invertebrates' digestive systems, on their excrement, and in layers coating the particles of organic material.

BACTERIA

Bacteria are responsible for most of the decomposition and heat generation in compost. They are the most nutritionally diverse group of compost organisms, using a broad range of enzymes to chemically break down a variety of organic materials. Bacteria are single-celled and are structured as rod-shaped bacilli, sphere-shaped cocci, or spiral-shaped spirilli. Many are motile, meaning they have the ability to move under their own power.

At the beginning of the composting process (up to 40°C), mesophilic bacteria predominate (Figure 1–8). They include hydrogen-oxidizing, sulfur-oxidizing, nitrifying, and nitrogen-fixing bacteria. Most of these can also be found in topsoil. Their populations increase exponentially during the initial stages of composting as they take advantage of the readily available simple compounds such as sugars and starches. Heat is produced by their metabolic activity, and if conditions are right, the compost begins to get hot.

Figure 1–8. Temperature Ranges for Compost Microorganisms.

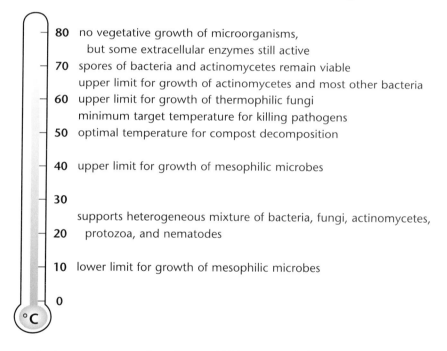

As temperatures rise above 40°C, mesophilic bacteria no longer thrive and thermophilic species take over. The microbial populations during this phase are dominated by members of the genus *Bacillus*. The diversity of bacilli species is fairly high at temperatures from 50–55°C but decreases dramatically above 60°C. When conditions become unfavorable, bacilli form thick-walled endospores that are highly resistant to heat, cold, and dryness. These spores are ubiquitous in nature and become active whenever environmental conditions are favorable.

At the highest compost temperatures, bacteria of the genus *Thermus* have been isolated. Composters sometimes wonder how microorganisms evolved in nature that can withstand the high temperatures found in active compost. *Thermus* bacteria were first found in hot springs in Yellowstone National Park, and they may have evolved there.[3] Other places where thermophilic conditions exist in nature include deep-sea thermal vents, manure droppings, and accumulations of decomposing vegetation.

Eventually, the compounds that are usable by thermophilic bacteria become depleted. As the activity of the thermophilic bacteria declines, the temperature falls and mesophilic bacteria again predominate. The numbers and types of mesophilic microbes that recolonize compost as it matures depend on what spores and organisms are present in the compost and the immediate environment. As the curing or maturation phase progresses, the diversity of the microbial community gradually increases. Eventually, the available carbon in the compost becomes depleted, and microbial populations once again drop.

ACTINOMYCETES

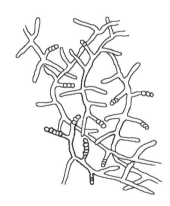

The characteristic earthy smell of soil is caused by actinomycetes, a type of bacteria that forms chains or filaments. In composting, actinomycetes play an important role in degrading complex organic molecules such as cellulose, lignin, chitin, and proteins. Although they do not compete well for the simple carbohydrates that are plentiful in the initial stages of composting, their enzymes enable them to chemically break down resistant debris, such as woody stems, bark, and newspaper, that are relatively unavailable to most other forms of bacteria and fungi. Some species of actinomycetes appear during the thermophilic phase, and others become important during the cooler curing phase, when only the most resistant compounds remain. Actinomycetes thrive under warm, well-aerated conditions and neutral or slightly alkaline pH.

Actinomycetes form long, threadlike branched filaments that look like gray spider webs stretching through compost. These filaments are most commonly seen toward the end of the composting process, in the outer 10 to 15 cm of the pile. Sometimes they appear as circular colonies that gradually expand in diameter.

FUNGI

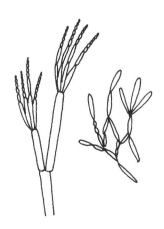

Fungi include molds and yeasts, and they are responsible for the decomposition of many complex plant polymers in soil and compost. In compost, fungi are important because they break down tough debris including cellulose. They can attack organic residues that are too dry, acidic, or low in nitrogen for bacterial decomposition. Most fungi secrete digestive enzymes onto the food, and then they absorb the products of extracellular digestion.

Fungal species are predominantly mesophilic. When temperatures are high, most are confined to the outer layers of compost. Compost molds are strict aerobes. They can be microscopic or appear as gray or white fuzzy colonies that are visible on the compost surface. Some fungi form chains of cells called hyphae that look like threads weaving through the organic matter. The mushrooms that you may find growing on compost are the fruiting bodies of some types of fungi. Each is connected to an extensive network of hyphae that reaches through the compost and aids in decomposition.

Research Possibility: *Composting recipes sometimes call for inoculating the pile by mixing in a few handfuls of finished compost. Is there any observable difference in appearance of microbes between systems that have and have not been so inoculated?*

PROTOZOA

Protozoa are one-celled microscopic organisms. In compost piles, they feed on bacteria and fungi. Protozoa make up only a small proportion of microbial biomass in compost.

INVERTEBRATES

Composting can occur either with or without the aid of invertebrates. In indoor commercial or industrial composting, invertebrates are often purposely excluded, and the systems are managed to promote thermophilic composting by microorganisms. Invertebrates are not active at the high temperatures that occur in thermophilic composting.

In contrast, compost in outdoor piles or bins provides an ideal habitat for a vast array of invertebrates commonly found in soil and decaying vegetation. Although most of the decomposition still is carried out by microorganisms, invertebrates aid in the process by shredding organic matter and changing its chemical form through digestion. If the compost heats up, the invertebrates may go into a dormant stage or move to the periphery of the pile where the temperatures are cooler.

Scientists use a number of systems for categorizing organisms that live in soil and compost. Different classification schemes provide different "filters" through which we view complex biological communities. The food web is one classification system, based on groups of organisms occupying the same trophic level (see Figure 1–6). Another way to classify compost invertebrates is by size (Figure 1–9). Body length sometimes is used to divide organisms into microfauna (<0.2 mm), mesofauna

Figure 1–9. Classification of Compost Organisms According to Body Width.

MICROFAUNA live in water films	MESOFAUNA live in air spaces	MACROFAUNA create space by burrowing
← < 0.1 mm →	← 0.1–2 mm →	← 2–60 mm →
Nematodes Protozoa	Mites Pseudoscorpions Springtails Potworms Flies	Earwigs Sowbugs Centipedes Millipedes Earthworms Slugs & Snails

(0.2–10 mm), and macrofauna (>10 mm). A similar classification scheme is based on body width. Body width is important because it specifies which organisms are small enough to live in the film of water surrounding compost particles, which live in the air-filled pore spaces, and which are large enough to create their own spaces by burrowing.

The commonly used taxonomic classification system that divides organisms into kingdom, phylum, class, order, family, genus, and species is based on phylogenic (evolutionary history) relationships among organisms. The following descriptions of common invertebrates found in compost are organized roughly in order of increasing size within the broad phylogenic classifications (Figure 1–10).

Figure 1–10. Phylogenic Classifications of Common Compost Organisms.

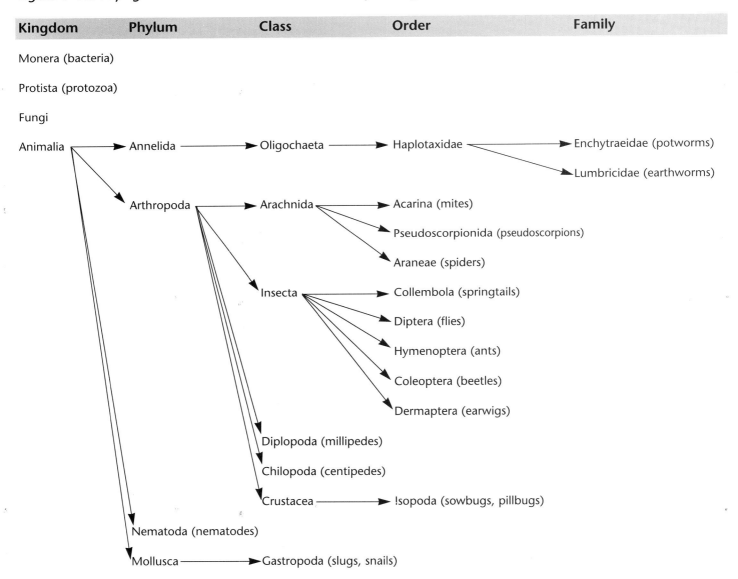

ANNELIDS

OLIGOCHAETES

Potworms (Phylum Annelida, Class Oligochaeta, Order Haplotoxidae, Family Enchytraeidae): Enchytraeids are small (10–25 mm long) segmented worms also known as white worms or potworms. Because they lack hemoglobin, they are white and can thus be distinguished from newly hatched, pink earthworms. Potworms often are found in worm bins and damp compost piles. They feed on mycelia, the thread-like strands produced by fungi. They also eat decomposing vegetation along with its accompanying bacterial populations.

Earthworms (Phylum Annelida, Class Oligochaeta): Because earthworms are key players in vermicomposting, they are described in greater detail later in this chapter (pp. 22–26).

ARTHROPODS

ARACHNIDS

Mites (Phylum Arthropoda, Class Arachnida, Order Acarina): There are over 30,000 species of mites worldwide, living in every conceivable habitat. Some are so specialized that they live only on one other species of organism. Like spiders, they have eight legs. They range in size from microscopic to the size of a pin head. Sometimes mites can be seen holding onto larger invertebrates such as sowbugs, millipedes, or beetles. Mites are extremely numerous in compost, and they are found at all levels of the compost food web (see Figure 1–6). Some are primary consumers that eat organic debris such as leaves and rotten wood. Others are at the secondary level, eating fungi or bacteria that break down organic matter. Still others are predators, preying on nematodes, eggs, insect larvae, springtails, and other mites.

Pseudoscorpions (Phylum Arthropoda, Class Arachnida, Order Pseudoscorpionida): Pseudoscorpions look like tiny scorpions with large claws relative to their body size, but lacking tails and stingers. They range from one to several millimeters in size. Their prey includes nematodes, mites, springtails, and small larvae and worms. Lacking eyes and ears, pseudoscorpions locate their prey by sensing odors or vibrations. They seize victims with their front claws, then inject poison from glands located at the tips of the claws. A good way to find pseudoscorpions is by peeling apart layers of damp leaves in a compost pile.

Spiders (Phylum Arthropoda, Class Arachnida, Order Araneae): Spiders feed on insects and other small invertebrates in compost piles.

INSECTS

Springtails (Phylum Arthropoda, Class Insecta, Order Collembola): Springtails are small, wingless insects that are numerous in compost. A tiny spring-like lever at the base of the abdomen catapults them into the air when they are disturbed. If you pull apart layers of decaying leaves, you are likely to see springtails hopping or scurry-

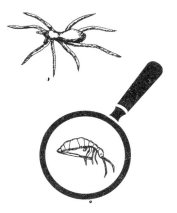

ing for cover. They feed primarily on fungi, although some species eat nematodes or detritus.

Flies (Phylum Arthropoda, Class Insecta, Order Diptera): Flies spend their larval phase in compost as maggots, which do not survive thermophilic temperatures. Adults are attracted to fresh or rotting food, and they can become a nuisance around worm bins or compost piles if the food scraps are not well covered. Fruit flies and fungus gnats, both of which can become pests in poorly managed compost piles, are in this order.

Ants (Phylum Arthropoda, Class Insecta, Order Hymenoptera): Ants eat a wide range of foods, including fungi, food scraps, other insects, and seeds. Ant colonies often can be found in compost piles during the curing stage.

Beetles (Phylum Arthropoda, Class Insecta, Order Coleoptera): The most common beetles in compost are the rove beetle, ground beetle, and feather-winged beetle. Feather-winged beetles feed on fungal spores, while the larger rove and ground beetles prey on other insects, snails, slugs, and other small animals.

Earwigs (Phylum Arthropoda, Class Insecta, Order Dermaptera): Earwigs are distinguished by jaw-like pincers on the tail end. Some species are predators and others eat detritus. They are usually 2–3 cm long.

OTHER ARTHROPODS

Millipedes (Phylum Arthropoda, Class Diplopoda): Millipedes have long, cylindrical, segmented bodies, with two pairs of legs per segment. They are slow moving and feed mainly on decaying vegetation. Stink glands along the sides of their bodies provide some protection from predators.

Centipedes (Phylum Arthropoda, Class Chilopoda): Centipedes can be distinguished from millipedes by their flattened bodies and single pair of legs per body segment. They are fast-moving predators found mostly in the surface layers of the compost heap. Their formidable claws possess poison glands used for paralyzing small worms, insect larvae, and adult arthropods including insects and spiders.

CRUSTACEANS

Sowbugs and Pillbugs (Phylum Arthropoda, Class Crustacea, Order Isopoda): Sowbugs, also called isopods, potato bugs, or wood lice, are the only terrestrial crustacean. Because they lack the waxy cuticle common to most insects, they must remain in damp habitats. They move slowly, grazing on decaying wood and resistant tissues such as the veins of leaves. Pillbugs, or rollypolies, are similar to sowbugs, except they roll into a ball when disturbed, whereas sowbugs remain flat.

NEMATODES

Nematodes (Phylum Nematoda): Under a magnifying lens, nematodes, or roundworms, resemble fine human hair. They are cylindrical and often transparent. Nematodes are the most abundant of the invertebrate decomposers—a handful of decaying compost probably contains several million. They live in water-filled pores and in the thin films of water surrounding compost particles. Some species scavenge decaying vegetation, some eat bacteria or fungi, and others prey on protozoa and other nematodes.

MOLLUSKS

Slugs and Snails (Phylum Mollusca, Class Gastropoda): Some species of slugs and snails eat living plant material, whereas others feed on decaying vegetation. Unlike many other invertebrates, some snails and slugs secrete cellulose-digesting enzymes rather than depending on bacteria to carry out this digestion for them.

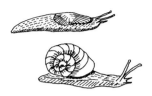

EARTHWORMS[4]

It may be doubted whether there are many other animals which have played so important a part in the history of the world, as have these lowly organized creatures.

Charles Darwin 1881[5]

If your only encounter with earthworms has been those shriveled-up specimens that didn't make it back to the grass after a rainstorm, you may not have the same appreciation for these lowly creatures that Charles Darwin did. But if you are a backyard or worm composter, you might have become fascinated by these burrowing invertebrates, and even have some questions about their role in the compost pile. What are they actually eating, and what comes out the other end? Are worm castings finished compost, or do they get broken down further? How do worms interact with compost microbes during decomposition? Can worm compost enhance the growth of plants?

To answer these questions, it is important to understand that if you've seen one worm, you definitely haven't seen them all. When aquatic forms are included, there are about 3,000 species of earthworms or members of the class Oligochaeta worldwide. Among these species, there is great variety in size, ranging from 10 mm–1.2 m in length and 10 mg–600 g in weight. Earthworms also exhibit diverse eating habits and ecological and behavioral characteristics. Thus, the answers to the above questions are complicated; what is true for one species is not necessarily true for another. Furthermore, because scientific research has been limited to only about 5% of the total number of worm species, we do not know the answers to many questions for most species of worms.

The information presented below is a synthesis based on many scientific experiments, though there could be exceptions to some of the generalizations for any one species. We also include information specific to the Lumbricidae, which is one of the most important earthworm families in terms of human welfare, and to one of its members, *Eisenia fetida*, the species most commonly used in vermicomposting.

EARTHWORM FEEDING AND DECOMPOSITION

Scientists have used several methods to determine the role of worms and other invertebrates in decomposition. In one experiment, organic materials of known weight were placed in mesh bags with different-size holes. The bags were then buried in soil. Several months to a year later, the scientists dug out the bags and determined the dry weight of the remaining organic material. It turned out that more decomposition had occurred in bags with holes large enough for earthworms than in those that allowed only smaller invertebrates access to the organic materials.

Through these and similar experiments, researchers have determined that much organic matter, particularly the tougher plant leaves, stems, and root material, breaks down more readily after being eaten by soil invertebrates. And, of all the invertebrates who play a role in the initial stage of organic matter decomposition, earthworms are probably the most important.

Research Possibility: Do organic wastes in compost break down more readily in the presence of worms than through composting that depends solely on microbial decomposition?

Worms that are active in compost are detritivores, feeding primarily on relatively undecomposed plant material. Some species live and feed in the upper organic or litter layer of soil. *Eisenia fetida*, the species most commonly used for vermicomposting, is one example of this type of worm. Other species live in deep soil burrows and come to the surface to feed on plant residues in the litter layer. By pulling leaves and other food down into their burrows, they mix large amounts of organic matter into the soil. *Lumbricus terrestris*, a worm commonly seen in North American gardens, is typical of this group. A third type of worm is not commonly seen in gardens or compost piles because it burrows deep beneath the surface and ingests large quantities of soil containing more highly decomposed plant material.

Anyone who has ever observed earthworm castings will recognize that they contain organic particles that are reduced in size relative to the leaves or other organic matter that the worms ingest. Organic matter passing through a worm gut is transformed chemically as well as physically. However, most worms are able to digest only simple organic compounds such as sugars. A few species, including *Eisenia fetida*, apparently are able to digest cellulose. No species has been found that breaks down lignin.

Worms both influence and depend on microbial populations in soil and compost. They feed on soil microorganisms, including fungi, bacteria, protozoa, amoebae, and nematodes. These organisms are probably a major source of nutrients for worms. Preferential feeding on different microorganisms may alter the microbial populations in soil or compost. Worms have also developed a symbiotic relationship with microbes inhabiting their digestive tract. The mucus found in the worm's intestine provides a favorable substrate for microorganisms, which in turn decompose complex organic compounds into simpler substances that are digestible by the worm. Some of the worm's mucus is excreted along with the casts, and it continues to stimulate microbial growth and activity in the soil or compost. The high levels of ammonia and partially decomposed organic matter in casts provide a favorable substrate for microbial growth. Thus, fresh worm casts generally have high levels of microbial activity and high decomposition rates. This activity decreases rapidly over a period of several weeks as degradable organic matter becomes depleted.

Research Possibility: How do respiration rates, a measure of microbial and invertebrate activity, vary in worm compost and other composts over time? How do chemical and physical properties differ between worm and other composts?

WORMS AND PLANT GROWTH

Nutrients are transformed during their passage through the worm gut into forms more readily available to plants, such as nitrate, ammonium, biologically available phosphorus, and soluble potassium, calcium, and magnesium. Because of these and other changes in soil and organic matter chemistry, physical properties, and biology brought about by worms, plants generally grow faster in soils with worms than in soils without them. Furthermore, studies have shown that extracts from worm tissues enhance plant growth.

Research Possibility: Much of the research on vermicompost and plant growth has been conducted with worms grown on sewage sludge. Does worm compost produced from other organic materials enhance plant growth? How do vermicomposts produced from various organic materials differ in terms of nutrient content? How do vermicomposts compare to composts produced by microbial activity alone? To composts that have a diversity of soil invertebrates?

Vermicompost is a finely divided material that has the appearance and many of the characteristics of peat. In some studies, it has been shown to enhance soil structure, porosity, aeration, drainage, and moisture-holding capacity. Its nutrient content varies depending on the original organic materials. However, when compared with a commercial plant-growth medium to which inorganic nutrients have been added, vermicompost usually contains higher levels of most mineral elements, with the exception of magnesium. It has a pH of about 7.0, and because most plants prefer slightly acidic conditions, vermicompost should be acidified or mixed with a more acid material such as peat, prior to use as a growth medium. Another adjustment sometimes made when using vermicompost for plant growth is to add magnesium. Because vermicomposting does not achieve high temperatures, sometimes a thermophilic stage is used prior to adding worms to kill insects and pathogens.

Research Possibility: In some experiments, plants did not show increased growth when planted in fresh worm castings. Does aging or "curing" worm castings increase their ability to enhance plant growth? Are there chemical differences between fresh and older worm castings? Should worm compost be mixed with soil before being used to grow plants?

Research Possibility: In China, farmers dig parallel trenches and fill them with organic wastes mixed with cocoons of Eisenia fetida. *Soybeans planted in rows between the trenches are highly productive. Can you design and test a planting system using vermicompost?*

EARTHWORMS AND WATER

Earthworms require large amounts of water, which they ingest with food and absorb through their body walls. The water is used to maintain a moist body surface that aids the worm's movement through soil and protects it against toxic substances. A moist body surface is also necessary because worms obtain oxygen by absorption in solution through their cuticle. A soil moisture content of 80–90% by weight is considered optimal for *Eisenia fetida*. Many worms, including *Lumbricus terrestris*, can tolerate poorly ventilated soils because of the high affinity of their hemoglobin for oxygen. However, under saturated soil conditions, worms will come to the surface, sometimes migrating considerable distances. It is unknown whether low oxygen levels or chemicals in soil solution cause this behavior.

Because most water uptake and loss occurs through the thin permeable cuticle, worms are at constant risk of dehydration. Although worms have no shell or waxy cuticle to maintain body moisture, they can survive low moisture conditions. Some species migrate to deeper soil levels when surface soil dries out.

LUMBRICIDAE

The family Lumbricidae is the dominant family of worms in Europe. As European agricultural practices spread throughout much of the world, so did Lumbricid worms. These worms were able to successfully colonize new soils and became dominant, often replacing endemic worm species. They are now the dominant family in most temperate, crop-growing regions around the world, including North America.

Lumbricus terrestris is one of the most common earthworms in northern North America. It lives in a wide range of habitats, including grasslands, agricultural fields, gardens, and forests. It feeds on leaves and other plant materials, dragging them into its burrows in the soil.

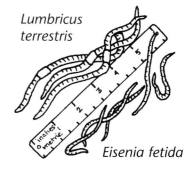

Lumbricus terrestris

Eisenia fetida

Eisenia fetida is the favored species for use in vermicomposting. It is particularly well suited to composting because it is extremely prolific, thrives in high organic matter habitats, can tolerate a wide range of temperatures and moisture conditions, and can be readily handled. Its natural habitat is probably under the bark of dead tree trunks, but it is most commonly found in animal dung, compost, and other accumulations of decaying plant material. Originally from Europe, it has become established throughout much of the world.

Just how prolific is *Eisenia fetida*? When it is given high-energy and nitrogen-rich food (such as horse manure or activated sewage sludge—yum!), adequate moisture, and optimum temperatures (25°C), cocoon production in *Eisenia fetida* starts 35 days after the worms hatch, and it reaches its maximum at 70 days. The cocoons, each carrying one to six eggs, are produced in the clitellum, or swollen region along the worm's body. Between three and four cocoons are produced each week. Nineteen days later, the young worms or hatchlings emerge and the process begins again. Thus, under "luxury" conditions, a population of *Eisenia fetida* can have four generations and produce 100 times its own weight in one year. Low food quality, overcrowding, or suboptimal temperatures or moisture levels reduce these reproductive rates. For example, growth is 24 times faster at 25°C than at 10°C, and temperatures below 0°C and over 35°C are considered lethal.

Research Possibility: *How do different food sources affect reproductive and growth rates of* Eisenia fetida?

Cocoons of *Eisenia fetida* may survive dryness and possibly other adverse conditions for several years and then hatch when favorable conditions return. In compost piles, adults may move from areas of less favorable conditions to areas with conditions conducive to their growth. For example, in winter months, they may migrate to the warm center of a large outdoor pile. Perhaps through these "mini-migrations" or through cocoon survival, *Eisenia fetida* are able to survive winters in regions where temperatures dip well below those that are lethal in the laboratory.

Research Possibility: *Eisenia fetida does best in wastes with pH between 5.0 and 8.0. How sensitive are cocoons to pH? Will they hatch after being exposed to extreme pH? How sensitive are they to extreme drought or temperatures?*

Does the use of *Eisenia fetida* in composting serve to further spread this exotic species, possibly interfering with native earthworm populations? To answer this question, it is useful to consider the fact that populations of earthworms are already much altered throughout the globe. In North America, for example, there are 147 species of worms, 45 of which were probably introduced. In fact, when Europeans first arrived in formerly glaciated parts of North America, they claimed there were no earthworms present. (It is assumed that earthworms in northern North America were wiped out during glaciation.) Thus, the species that are currently in these regions were either introduced in soil from imported plants or spread northward from southern regions of North America. *Eisenia fetida* is thought to have been introduced to North America in organic soils brought in with imported plants. Because it is adapted to compost and other organic substrates, it is unlikely to spread into neighboring soils and compete with soil-inhabiting worms.

---

[1] Kayhanian, M. and G. Tchobanoglous. 1992. Computation of C/N ratios for various organic fractions. *BioCycle* 33(5): 58–60.

[2] Michel, F. C., Jr., L. J. Forney, A. J.-F. Huang, S. Drew, M. Czuprenski, J. D. Lindeberg, and C. A. Reddy. 1996. Effects of turning frequency, leaves to grass mix ratio and windrow vs. pile configuration on the composting of yard trimmings. *Compost Science & Utilization* 4: 26–43.

[3] F. C. Michel, Jr., NSF Center for Microbial Ecology, Michigan State University (personal communication).

[4] The information on earthworms was compiled from the following sources:

Edwards, C. A. and P. J. Bohlen. 1996. *Biology and Ecology of Earthworms*. Chapman Hall. London, U.K.

Edwards, C. A. and E. F. Neuhauser, editors. 1988. *Earthworms in Waste and Environmental Management*. SPB Academic Publishing. The Hague, The Netherlands.

Hendrix, P. F., editor. 1995. *Earthworm Ecology and Biogeography in North America*. Lewis Publishers, CRC Press. Boca Raton, FL.

Lee, K. E. 1985. *Earthworms: Their Ecology and Relationships with Soils and Land Use*. Academic Press, Australia.

[5] Darwin, C. 1881. *The Formation of Vegetable Mould Through the Action of Worms With Observations on Their Habits*. Currently published by Bookworm Publishing Co., Ontario, CA.

2
COMPOSTING BIOREACTORS AND BINS

When you think of composting, chances are you envision one of a variety of bins that are used for composting outdoors. But composting can also be carried out right in the classroom, using containers or bioreactors that range in size from soda bottles to garbage cans. Although these bioreactors are generally too small to recycle large quantities of organic wastes, they are ideal for conducting composting research and designing small-scale models of larger composting systems. Because these units are small, they need to be carefully designed and monitored to provide conditions favorable for aerobic, heat-producing composting. Two types of bioreactors for indoor thermophilic composting are described in this chapter:

- **TWO-CAN BIOREACTORS** are made from nested garbage cans. Each two-can system will process enough organic matter to fill a 20-gallon can. Temperature increases indicating microbial activity can be observed within the first week of composting, and the compost should be finished and ready for curing within two to three months.

- **SODA BOTTLE BIOREACTORS** are used as tools for research rather than waste management. They are small and inexpensive, enabling students to design and carry out individualized research projects comparing the effect of variables such as reactor design, moisture content, and insulation on compost temperature.

Another option for indoor composting is to use worm bins, which rely on red worms (commonly *Eisenia fetida*) working in conjunction with mesophilic microbes to decompose food scraps and bedding materials, producing finished compost in three to five months. Vermicompost systems provide many opportunities for the study of worm behavior.

We also include recommendations for bins that can be used outdoors in the school yard. The intricate food web that is created by soil invertebrates invading an outdoor compost pile provides an ideal focus for biological and ecological studies. Outdoor compost systems generally are larger than indoor units and can be used to recycle sizable quantities of food and yard wastes. In piles or bins in which thermophilic composting occurs, compost can be produced within a couple of months. If the compost does not get hot, a year or longer will be needed before the materials have broken down.

Students can use the designs included in this chapter, or develop their own composting systems based on their own knowledge and creativity.

Research Possibility: How does the type of composting system (bin, bioreactor, pile) affect the heat produced, the time it takes to produce compost, or the quality of the end product?

TWO-CAN BIOREACTORS

PURPOSE

Two-can composters consist of a 20-gallon garbage can containing organic wastes placed inside a 32-gallon garbage can. Although many classrooms have successfully composted with a single container, placing the can that holds wastes inside another container helps alleviate any odor and fly problems that may arise. The outside container can also be used to collect leachate.

Two-can units are designed to be used for small-scale indoor composting, and as an educational tool in the classroom. A 20-gallon can holds only about 10% of the cubic meter volume commonly recommended for thermophilic composting. Thermophilic composting is possible in these smaller systems, but careful attention needs to be paid to C:N ratios, moisture content, and aeration.

A system using a 10-gallon plastic garbage can inside a 20-gallon can may be substituted if space is a problem. The smaller system may operate at lower temperatures, thereby lengthening the time for decomposition. Or students may want to experiment with various aeration and insulation systems to see if they can come up with a 10-gallon system that achieves temperatures as high as those in a larger system.

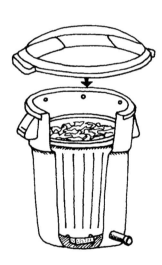

MATERIALS

- 32-gal plastic garbage can
- 20-gal plastic garbage can
- drill
- brick
- spigot (optional—see Step 3, below)
- insulation (optional—see Step 5, below)
- duct tape (optional—see Step 5, below)
- 6 pieces of nylon window screen (each about 5 cm^2)
- dial thermometer with stem at least 60 cm long
- peat moss or finished compost to make 5-cm layer in outer can
- compost ingredients, including high-carbon materials such as wood chips and high-nitrogen materials such as food scraps (see Step 8, below)

CONSTRUCTION

1. Using a drill, make 15 to 20 holes (1–2.5 cm diameter) through the bottom of the 20-gal can.

2. Drill five 1–2.5 cm holes just below the rim of the larger garbage can, and cover them on the inside with pieces of nylon window screen.

3. Design and build a spigot at the bottom of the larger can for draining leachate. One way to do this is to fit a piece of pipe into a hole at the bottom edge of the outer can, sealing around the edges with waterproof tape or sealant. Close the outer end of the pipe with a tight-fitting cork or stopper that can be removed to drain the accumu-

lated leachate, and cover the inner end with a piece of nylon screening to block the flow of solid particles.

4. Place a brick or some other object in the bottom of the 32-gal can. This is to separate the two cans, leaving space for leachate to collect. (Students may want to measure the leachate and add it back into the compost.)

5. If you are composting in a cold area, you may want to attach insulation to the outer barrel and lid with duct tape, making sure not to block aeration holes.

6. To reduce potential odors, line the bottom of the outer can with several centimeters of absorbent material such as peat moss or finished compost. Periodically drain the leachate to avoid anaerobic conditions that may cause odors. The leachate can be poured back in the top if the compost appears to be drying out. Otherwise, dispose of it outside or down the drain, but do not use it for watering plants. (This leachate is not the "compost tea" prized by gardeners, and it could harm vegetation unless diluted. Compost tea is made by soaking mature compost, after decomposition is completed.)

7. Fill the reactor, starting with a 5–10 cm layer of "brown" material such as wood chips, finished compost, or twigs and branches. Loading can take place all at once (called "batch composting") or in periodic increments. With batch composting, you are more likely to achieve high temperatures quickly, but you will need to have all organic material ready to add at one time. If you are going to add layers of materials over a period of time rather than all at once, the material probably won't begin to get hot until the can is at least 1/3 full (Figure 2–1).

Figure 2–1. A Typical Temperature Profile for Two-Can Composting with Continuous Loading.

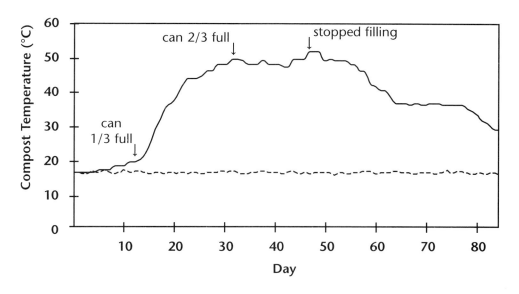

Whether you fill the reactor all at once or in batches, remember to keep the ingredients loose and fluffy. Although they will become more compact during composting, never pack them down yourself because the air spaces are needed for maintaining aerobic conditions. Another important rule is to keep the mixture in the inner can covered at all times with a layer of high-carbon material such as finished compost, sawdust, straw, or wood shavings. This minimizes the chance of odor or insect problems.

8. To achieve thermophilic composting, you will need to provide the ingredients within the target ranges for moisture, carbon, and nitrogen. For moisture, the ideal mixture is 50–60% water by weight. You can calculate this by using the procedure described in Chapter 3 (pp. 47–48), or use the rule of thumb that the ingredient mix should feel about as damp as a wrung-out sponge. For carbon and nitrogen, the mixture should contain approximately 30 times as much available carbon as nitrogen (or a C:N ratio of 30:1). Using a specified quantity of one ingredient, you can calculate how much of the other you will need to achieve this ratio (see Chapter 3, pp. 48–50). Or, you can simply make a mixture of high-carbon and high-nitrogen materials. Organic materials that are high in carbon include wood chips or shavings, shredded newspaper, paper egg cartons, and brown leaves. Those high in nitrogen include food scraps, green grass or yard trimmings, coffee grounds, and manure. (Do not use feces from cats or meat-eating animals because of the potential for spreading disease organisms.)

You are now ready to begin monitoring the composting process using the methods outlined in Chapter 4. The composting process should take two to three months after the can is filled. At the end of this period, you can either leave the compost in the can or transfer it into other containers or an outdoor pile for the curing phase.

SODA BOTTLE BIOREACTORS

PURPOSE

Soda bottle bioreactors are designed to be used as tools for composting research rather than as a means to dispose of organic waste. They are small and inexpensive, enabling students to design and carry out individualized research projects comparing the effect of variables such as moisture content or nutrient ratios on compost temperatures.

Use the instructions below as a starting point. Challenge students to design their own soda bottle reactors and to monitor the temperatures that their reactors achieve.

MATERIALS

- two 2-liter or 3-liter soda bottles
- Styrofoam plate or tray
- one smaller plastic container such as a margarine tub that fits inside the bottom of the soda bottle (optional—see Step 3, below)
- drill or nail for making holes
- duct tape or clear packaging tape
- utility knife or sharp-pointed scissors
- insulation materials such as sheets of foam rubber or fiberglass
- fine-meshed screen or fabric (such as a piece of nylon stocking) large enough to cover ventilation holes to keep flies out
- dial thermometer with stem at least 20 cm long
- chopped vegetable scraps such as lettuce leaves, carrot or potato peelings, and apple cores, or garden wastes such as weeds or grass clippings
- bulking agent such as wood shavings or 1-cm pieces of paper egg cartons, cardboard, or wood
- hollow flexible tubing to provide ventilation out the top (optional— see Step 8, below)

CONSTRUCTION

1. Using a utility knife or sharp-pointed scissors, cut the top off one soda bottle just below the shoulder and the other just above the shoulder. Using the larger pieces of the two bottles, you will now have a top from one that fits snugly over the bottom of the other.

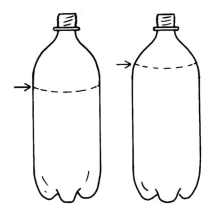

2. The next step is to make a Styrofoam circle. Trace a circle the diameter of the soda bottle on a Styrofoam plate and cut it out, forming a piece that fits snugly inside the soda bottle. Use a nail to punch holes through the Styrofoam for aeration. The circle will form a tray to hold up the compost in the bioreactor. Beneath this tray, there will be air space for ventilation and leachate collection.

3. If your soda bottle is indented at the bottom, the indentations may provide sufficient support for the Styrofoam circle. Otherwise, you will need to fashion a support. One technique is to place a smaller plastic

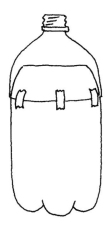

container upside down into the bottom of the soda bottle. Other possibilities include wiring or taping the tray in place.

4. Fit the Styrofoam circle into the soda bottle, roughly 4–5 cm from the bottom. Below this tray, make air holes in the sides of the soda bottle. This can be done with a drill or by carefully heating a nail and using it to melt holes through the plastic. If you are using a plastic container to hold up the Styrofoam tray, you may need to drill holes through the container as well. The object is to make sure that air will be able to enter the bioreactor, diffuse through the compost, and exit through the holes or tubing at the top.

   Avoid making holes in the very bottom of the bottle unless you plan to use a pan underneath it to collect whatever leachate may be generated during composting.

5. Next, determine what you will compost. A variety of ingredients will work, but in general you will want a mixture that is 50–60% water by weight and has approximately 30 times as much available carbon as nitrogen (a C:N ratio of 30:1). You can estimate moisture by using the rule of thumb that the mixture should feel as damp as a wrung-out sponge, or you can calculate optimal mixtures using the procedures in Chapter 3 (pp. 47–48).

   Similarly, mixtures that will achieve optimal C:N ratios can be either estimated or calculated. Materials that are high in carbon include wood chips or shavings, shredded newspaper, and brown leaves. High-nitrogen materials include food scraps, green grass or yard trimmings, and coffee grounds. By mixing materials from the high-carbon and high-nitrogen groups, you can achieve a successful mixture for thermophilic composting. Try to include more than just a couple of ingredients; mixtures containing a variety of materials are more likely than homogeneous ones to achieve hot temperatures in soda bottle bioreactors. To calculate rather than estimate the amounts needed, use the equations in Chapter 3 (pp. 48–50).

   The particle size of compost materials needs to be smaller in soda bottle bioreactors than in larger composting systems. In soda bottles, composting will proceed best if the materials are no larger than 1–2 cm.

6. Loosely fill your bioreactor. Remember that you want air to be able to diffuse through the pores in the compost, so keep your mix light and fluffy and do not pack it down.

7. Put the top piece of the soda bottle on and seal it in place with tape.

8. Cover the top hole with a piece of screen or nylon stocking held in place with a rubber band. Alternatively, if you are worried about potential odors, you can ventilate your bioreactor by running rubber tubing out the top. In this case, drill a hole through the screw-on soda bottle lid, insert tubing through the hole, and lead the tubing either out the window or into a ventilation hood.

9. If you think flies may become a problem, cover all air holes with a piece of nylon stocking or other fine-meshed fabric.

10. Insulate the bioreactor, making sure not to block the ventilation holes. (Because soda bottle bioreactors are much smaller than the typical compost pile, they will work best if insulated to retain the heat that is generated during decomposition.) You can experiment with various types and amounts of insulation.

Now you are ready to watch the composting process at work! You can chart the progress of your compost by taking temperature readings. Insert a thermometer down into the compost through the top of the soda bottle. For the first few days, the temperature readings should be taken at least daily, preferably more often. In these small systems, it is possible that temperatures will reach their peak in less than 24 hours. To avoid missing a possible early peak, use a max/min thermometer or a continuously recording temperature sensor, or have the students measure the temperatures frequently during the first few days.

Soda bottle reactors generally reach temperatures of 40–45°C, somewhat lower than temperatures achieved in larger composting systems (Figure 2–2). If conditions are not right, no noticeable heating will occur. Challenge your students to design systems that show temperature increases, and use their results as a starting point for a discussion of the various factors that affect microbial growth and decomposition (C:N ratios, moisture levels, air flow, size, and insulation).

Figure 2–2. A Typical Temperature Profile for Soda Bottle Composting.

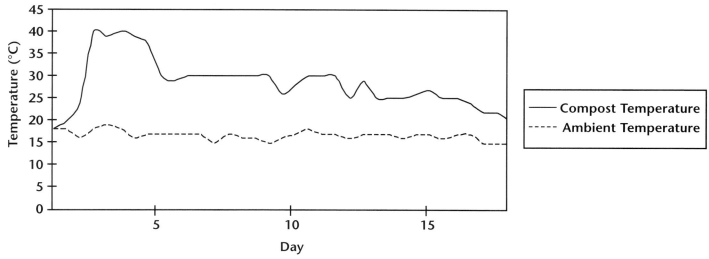

Research Possibility: *How do different soda bottle reactor designs affect the temperature profile during composting? How do different mixtures of organic materials affect the temperature profile in soda bottle reactors?*

Because soda bottles are so small, you may not end up with a product that looks as finished as the compost from larger piles or bioreactors. However, you should find that the volume shrinks by one-half to two-thirds and that the original materials are no longer recognizable. You can let the compost age in the soda bottles for several months, or transfer it to other containers or outdoor piles for curing.

WORM BINS

PURPOSE

Thousands of schools around the world use worm bins to teach students about recycling organic wastes and to involve them in investigations of worm biology and behavior. Vermicomposting (from the Latin *vermis*, for worm) is also used in large-scale industrial applications, with waste streams including sewage sludge, animal manure, and food wastes. Unlike thermophilic composting, vermicomposting does not get hot. In fact, temperatures above 35°C would kill the worms. In all types of composting, microorganisms play the key role in decomposition. In vermicomposting, worms help by physically breaking down the organic matter and chemically altering it through digestion. The end product, vermicompost, contains plant-available nutrients and compounds that enhance plant growth.

Vemicomposting provides a wealth of opportunities for student research on topics such as worm behavior, life cycles, feeding preferences, and effects of worms and other invertebrates on decomposition. For more information on the biology and ecology of worms, see Chapter 1 (pp. 22–26).

MATERIALS

- a worm bin (see Steps 1 and 2, below)
- bedding (see Step 3, below)
- red worms (see Step 4, below)
- food (see Step 5, below)

CONSTRUCTION

1. **Decide on the bin size:** If you plan to use your worm bin for classroom observation and scientific investigation, but not for recycling of a set amount of food scraps, then any size bin will do. It can be as small as a shoe box or as large as you'd like to make it. However, if you plan to put in a set amount of food, such as all the non-meat lunch scraps from your classroom, then you will need to calculate the size of the bin based on the amount of food you plan to compost. Conduct a waste audit by collecting food scraps for a week and then weighing how much has accumulated. The rule of thumb is that roughly a square foot of bin space is needed for every pound of waste composted per week. (Since red worms are surface dwellers, surface area rather than volume is used for this calculation).

2. **Prepare the bin:** Many different types of containers are successfully used for vermicomposting. Wooden boxes, Styrofoam coolers, and plastic tubs all are possibilities. Whatever type of container you use, providing adequate ventilation is the key to success. Worm bins usually have a tight-fitting lid, with many small ventilation holes drilled through the bin sides for ventilation. The most fail-proof design uses drain holes in the bottom, with the bin propped up on blocks so that excess moisture can drain into an underlying tray. (Most food scraps are wet, and without drainage the bedding can become mucky and

anaerobic.) If you want to build a self-contained unit, it is possible to maintain proper moisture conditions without bottom drainage, as long as you are willing to keep a watchful eye on the moisture level and mix in dry bedding whenever the compost mixture begins to look soggy. (You want the mix to be moist but without puddles.)

Some commercially available worm bins use a tiered system of stacking bins. When the worms have depleted the food in the bottom bin, they crawl upward through holes or mesh into the next higher bin, where fresh food scraps have been added. Students may want to design a system such as this, using a stack of mesh-bottomed trays or boxes.

3. **Prepare the bedding:** The bedding holds moisture and contains air spaces essential to worms, and it gets eaten along with the food wastes. Common types of bedding include strips of newspaper, office paper, corrugated cardboard, or paper egg cartons. Machine-shredded paper or cardboard is ideal if available. If not, paper or cardboard will need to be torn into strips 1–3 cm wide. Other popular bedding materials include leaves, sawdust, peat moss, and shredded coconut fiber, which is similar in consistency to peat moss and is available from companies that sell red worms.

    Whatever type of bedding you choose, you will need to soak it until saturated, then drain off any excess water. Place the damp bedding in the bin, filling it to a depth of about 20 cm. Do not pack the bedding down—leave it loose to provide air spaces for the worms. You may wish to add a couple of handfuls of sand or crushed egg shells to provide the grit that worms use to grind their food.

4. **Add the worms:** The species most commonly used for vermicomposting is *Eisenia fetida*, commonly called red worms, red wigglers, manure worms, or tiger worms. You will need a pound or two (1000–2000 worms) to get started, after which the worms will replace themselves as long as conditions remain suitable for their reproduction and growth. If you can't obtain *Eisenia fetida* from a composter or worm farm in your community, you can mail-order them.[1] *Lumbricus terrestris*, the common night crawlers found in gardens, will not thrive in worm bins because they are adapted to burrowing deep into soil. In contrast, *Eisenia fetida* are surface dwellers, adapted to living in organically rich surface soils and the overlying layers of decomposing leaves and organic debris.

5. **Add food:** Bury a few handfuls of fruit and vegetable scraps in the bedding. Wait several days for the worms to acclimatize, then gradually increase the amount of food based on how quickly it is disappearing. You can add food every day, or you can leave the bin untended for a week or even up to a month once it has become established.

    Any food that humans eat can be fed to worms, but some types are more suitable than others for indoor worm bins. Vegetable and fruit scraps are ideal, such as carrot peels, melon rinds, and apple cores. Coffee grounds, tea bags, bread crusts, and pasta all are suitable. If you add a large amount of citrus fruits or other acidic foods, it would be a good idea to monitor the pH to make sure it does not drop below the 6.5 to 8.5 range that is ideal for worms (see **pH**, below). It is best to limit the quantities of foods such as onions and broccoli, which tend to have strong odors as they decompose. Avoid meats, fats, and dairy products because these foods decompose slowly and may attract pests. You may want to cut or break large food scraps into small pieces for faster decomposition.

6. **Cover the bin** with a lid made of plastic, wood, or fabric to provide shade and conserve moisture.

7. **Locate the bin** in an area where it will not be exposed to extreme heat or cold. Red worms thrive at room temperature (15–25°C). They can survive in colder locations, but their rate of feeding will be reduced. They will die if they freeze or are subjected to temperatures above about 35°C. Avoid placing bins on surfaces that vibrate, such as the surface of washing machines, because the vibrations may trigger the worms to try to leave the bin.

MAINTENANCE

Maintaining a worm bin is simple. No mixing or turning is needed. All you need to do is to keep an eye on the amount of food, the moisture level, and possibly the pH, and to harvest the worms once the composting process is completed.

Food: Monitor the bin and add more food once the first scraps have started to disappear. How much food should you add? Red worms can eat up to half their weight per day, so a simple calculation based on their initial weight will provide a guideline for the appropriate amount of food. But, it is important not to rely on numbers alone—instead, check once a week or so to see if the food is disappearing, and adjust feeding levels accordingly.

Research Possibilities: Worm feeding provides many opportunities for experimentation. Which types of food disappear the fastest, and which decompose slowly? Which foods tend to attract large groups of worms, and which ones do the worms avoid? Does the initial size of food particles affect how quickly they disappear?

Moisture: If the bedding appears dry, add moist food scraps or lightly spray with water. If the bedding becomes soggy, add dry newspaper strips or other dry bedding material, and avoid adding wet food scraps until the moisture is back in balance.

pH: Although not essential, periodic pH measurements are useful in monitoring the conditions in your worm bin. Simply insert pH paper into damp vermicompost, or follow one of the other measurement techniques described on p. 54. *Eisenia fetida* do best at a pH around neutral or slightly alkaline, in the range of 6.5 to 8.5. If conditions become acidic, mix in a sprinkling of crushed egg shells or powdered lime ($CaCO_3$, *not* hydrated lime).

Bedding: Within several months, most of the original bedding will have been replaced with brown, soil-like castings. At this point, it is time to remove the completed vermicompost and harvest the worms to start a new batch. This can be done all at once or through a gradual process (see **Harvesting the Worms**, p. 39).

TROUBLESHOOTING

Escaping Worms: When you first put worms into fresh bedding, they may initially try to escape. If you keep the lid on, they will gradually get acclimatized and remain in the bin even when it is uncovered. Once your vermicomposting is underway, worms attempting to escape is a signal that conditions in the bin are not favorable. The bedding may be too moist, which you can solve by mixing in some dry bedding. Or, the conditions may have become too acidic (see **pH**, p. 37). Another possibility is that the worms are not getting enough to eat. If all of the food and bedding have decomposed, the remaining castings will not continue to adequately nourish the worms, and it is time to harvest them and begin again with fresh bedding.

Worm Mortality: In a properly maintained worm bin, worms will continually die and decompose without being noticed, and the population will replenish itself through reproduction. If you notice many dead worms, there is a problem. Either the mixture has become too wet or dry, too hot or cold, or available food supplies have become depleted (see remedies listed under **Escaping Worms**, above).

Fruit Flies: Although fruit flies do not pose any health hazards, they can be a nuisance. To avoid breeding flies in worm bins, make sure to bury all food scraps in bedding. Monitor the decomposition of food that you add, and hold off on adding more if the scraps are sitting longer than a few days before disappearing. Keep bedding material moist but not too wet, since overly wet conditions encourage the proliferation of fruit flies.

If a fruit fly problem does develop, stop adding food until the worms have had a chance to catch up with the existing supply, and add dry newspaper strips if the bedding appears soggy. If the outdoor weather is suitable, you might want to air out the bin by leaving it uncovered outside for a few hours.

You can build a simple but effective fruit fly trap that can be placed either right in the worm bin or anywhere in the classroom where flies congregate. Take a soda bottle and remove the lid. Cut the bottle in two, with the top part slightly shorter than the bottom part. Pour cider vinegar into the bottom part to a depth of about 2 cm. Then, invert the top of the bottle into the bottom, forming a funnel leading into the bottle. Fruit flies will be attracted to the vinegar, and they will drown or get trapped.

Other Invertebrates: Many types of soil invertebrates can inhabit worm bins without causing problems. If you use leaves for bedding or if you add soil for grit, you are likely to introduce into the bin a variety of invertebrates such as millipedes, sowbugs, slugs, and springtails (see pp. 19–21 for a description of common soil invertebrates). These organisms are decomposers and will aid in making compost. You may notice some tiny white worms in your vermicompost. These often are mistaken for baby red worms but can be distinguished by their white rather than pink coloration. They are potworms, or enchytraeids (see p. 19). Potworms will not harm red worms, and they will aid in decomposition, but if you have many of them in your bin it may indicate that conditions have become acidic. Mites usually are present in worm bins but rarely present a problem. However, if your vermicompost becomes too moist or acidic, it may become infested and appear to be swarming with mites. At this point, it is a good idea to harvest the worms, rinse them, and start over with new bedding and food in a clean bin.

Odors: A properly functioning worm bin will not have a noticeable odor. If unpleasant smells do develop, there are several possible reasons. The bedding may have become too wet or may not be getting enough air. Another possible source of odors is food such as onions or broccoli that are naturally smelly as they decompose, or foods such as dairy products that turn rancid because they are not eaten rapidly by worms. To correct a problem of this sort, simply remove and dispose of the offending foods.

HARVESTING THE WORMS

If you want to use the compost on plants, and reuse the worms to make more compost, you will need to harvest the worms. This is generally done after three or more months, when the bin is filled with compost and very little bedding remains. Two methods are commonly used to separate worms from the castings. In the slower but easier method, you simply push all of the worm bin contents to one-half of the bin. In the empty half, start a new batch of compost by providing fresh bedding and food scraps. Over the next several weeks, the worms will move to the side with new food, conveniently leaving their castings behind in the other section. After this occurs, you can remove the finished compost and replace it with fresh bedding.

A faster but more labor-intensive method of removing worms involves dumping the entire contents of the worm bin onto a sheet of plastic or paper in a sunny or brightly lit location. Shape the compost into several cone-shaped piles. The worms will burrow downward to avoid light. Scoop the top layer off each pile, wait a few minutes for the worms to burrow farther, and then remove the next layer of compost. Repeat this process until the worms have become concentrated at the bottom of each pile. Collect the worms, weigh them (to compare with your initial mass of worms), and put them back in the bin with fresh bedding.

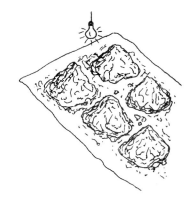

Research Possibility: *Design some experiments to find a better way of separating worms from compost. Are there any chemical or physical factors that will cause the worms to migrate, without impairing the beneficial properties of finished vermicompost?*

USING VERMICOMPOST

Once the food and bedding have fully decomposed, vermicompost can be used in the same way as other types of compost, either mixed into the soil or added to the surface as a top dressing. By mixing vermicompost with water, you can make a compost tea solution that enriches plants as they are watered. Many claims are made about the growth-inducing properties of vermicompost. See Chapters 6 and 7 for ideas about experiments to test the effects of various types of compost on plant growth.

Research Possibility: *How does vermicompost compare with thermophilic compost in plant growth experiments?*

OUTDOOR COMPOSTING

Outdoor composting systems can be larger than indoor bioreactors, allowing students to compost greater quantities of food scraps and landscaping trimmings. Although slightly less convenient than a system right in the classroom, students can monitor the temperature, moisture content, and other aspects of an outdoor system, and they can bring samples of the compost inside for observation and experimentation. Many schools have developed outdoor composting systems into demonstration sites, with signs explaining the composting process.

Unlike indoor systems, outdoor systems are home to a diverse range of invertebrates such as millipedes, centipedes, earthworms, pseudoscorpions, beetles, snails, mites, and springtails. These organisms form an intricate food web, and they can be used for illustrating ecological principles as well as for investigating topics such as life cycles and feeding preferences.

In some outdoor systems, the organic materials are periodically mixed or "turned." This redistributes materials that were on the outside of the pile and exposes them to the higher levels of moisture, warmth, and microbial activity found in the center. It also fluffs up the compost materials, allowing better air flow through the pile. The net result generally is to speed up the composting process (see pp. 10–11).

Bins should be located close to a water source in case they become too dry. Good drainage is also important in order to avoid standing water and the build-up of anaerobic conditions. Other considerations include avoiding exposure to high winds which may dry and cool the pile, and to direct sunlight which may also dry out the pile. The pile should not touch wooden structures or trees because it may cause them to decay. There should be space nearby for temporary storage of organic wastes.

There is an endless variety of outdoor composting systems. Bins may be purchased or constructed. Three types of systems are described below. Refer to *Composting: Wastes to Resources* (Bonhotal and Krasny, 1990) for more details on outdoor bin designs, or have the students design their own outdoor bins using readily available scrap materials.

HOLDING UNITS

Holding units provide a low-maintenance form of composting. You simply build the unit, fill it with organic materials, and then wait for the materials to decompose. A holding unit can be any container that holds organic materials while they are breaking down. The unit should be about a cubic meter in size (1 m x 1 m x 1 m), and it can be built from wire mesh, snow fence, cinder blocks, wooden pallets, or other materials. You can fill holding units with high-carbon materials such as autumn leaves and yard trimmings, realizing that these materials by themselves will not heat up and will require a year or more to fully decompose. If your system is dominated by leaves, you may want to avoid adding any food scraps, which might attract rodents or raccoons during the slow decomposition process. Alternatively, if you start with a mix that has the

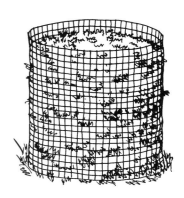

right C:N ratio and moisture level to become thermophilic, food scraps should break down quickly before any pests become a problem.

TURNING UNITS

A turning unit looks like three holding units placed side by side. Each unit should be a cubic meter (1 m x 1 m x 1 m) in size. Leave one side open or build a gate along one edge for easy access. Fill one bin at a time, using a mixture of high-nitrogen and high-carbon materials. For rapid

composting, turn the contents into the empty adjoining bin every week or two, or each time the temperature begins to decline. A pile that is kept "hot" like this should produce compost within a couple of months, although an additional period of curing is necessary before the compost is used for growing plants. The final bin provides the space needed for curing while a new batch of compost is started in the first bin.

ENCLOSED BINS

For small-scale outdoor composting, enclosed bins are an option. They can be purchased from home and garden centers or inexpensively built from a large garbage can. Simply drill 2-cm aeration holes in rows at roughly 15-cm intervals around the can. Fill the cans with a mixture of high-carbon and high-nitrogen materials. Stir the contents occasionally to avoid anaerobic pockets and to speed up the composting process. Although no type of bin is rodent-proof, enclosed bins do help to deter rodents and are popular for food scrap composting.

---

[1] Worms are available by mail from sources such as Beaver River Associates, PO Box 94, West Kingston, RI 02892 (401) 782–8747, or Flowerfield Enterprises, 10332 Shaver Rd., Kalamazoo, MI 49002 (616) 327–0108.

3
GETTING THE RIGHT MIX[1]

Once the bin or bioreactor is constructed, it is time to add the organic materials that will form compost. If you have several materials that you want to compost, how do you figure out the appropriate mix to ensure the proper balance of carbon, nitrogen, and moisture content? You may want to do this by simply estimating the right mix of materials that are wet and rich in nitrogen, such as food scraps, with other materials that soak up moisture and are high in carbon, such as wood chips. In fact, many experienced composters develop a "feel" for what works, following the general rules of thumb described in this chapter for estimating appropriate mixtures of compost ingredients.

Other composters, especially compost researchers and operators of large-scale commercial systems, combine general guidelines with more precise calculations. For those who prefer to be exact about adding ingredients, the second section of this chapter shows how to calculate a mixture of organic materials that is balanced in terms of carbon, nitrogen, and moisture content.

CHOOSING THE INGREDIENTS: GENERAL RULES OF THUMB

MOISTURE

One of the most important considerations for successful composting is the moisture content of the ingredients. In general, you want to achieve a balance between materials high in moisture, such as fruit and vegetable scraps, with dry materials such as wood chips. A common rule of thumb is that the compost mixture has the right moisture content if it is about as wet as a wrung-out sponge, with only a drop or two expelled when it is squeezed.

Composting proceeds best at a moisture content of 50–60% by weight. Table 3–1 lists typical moisture contents of common compost ingredients.

Table 3–1. Moisture Content of Common Compost Ingredients.[2] *(These data should be viewed as representative ranges, not as universal values.)*

Material	Moisture content (% wet weight)
Vegetables and fruits	80–90
Grass clippings	80
Leaves	40
Sawdust	40
Shrub trimmings	15

Instead of relying on typical figures such as those in Table 3–1, you may wish to measure the moisture content of your compost. You can use the following procedure to measure moisture either in the mixture as a whole or in the individual ingredients prior to mixing.

> **MEASURING COMPOST MOISTURE**
>
> 1. Weigh a small container, such as a paper plate or cupcake wrapper.
>
> 2. Weigh 10 g of compost into the container.
>
> 3. Dry the sample for 24 hours in a 105–110°C oven. (Using a microwave oven is a possible alternative, but this requires some experimentation to determine the drying time. Begin by spreading the compost into a thin layer in a microwave-safe container, then heat for one minute at full power. Remove the sample and weigh it. Reheat for another minute, then weigh it again. Repeat this cycle until the weight change becomes negligible. If the sample becomes burned or charred, start over and use reduced power or shorter heating times.)
>
> 4. After drying, reweigh the sample, subtract the weight of the container, then calculate the percent moisture content using the following equation:
>
> *Equation 1:*
>
> $$M = \frac{W_w - W_d}{W_w} \times 100$$
>
> *in which:*
> M = *moisture content (%) of compost sample*
> W_w = *wet weight of the sample, and*
> W_d = *weight of the sample after drying.*
>
> If you have used a 10 g compost sample, this simplifies to:
>
> $$M = \frac{10 - W_d}{10} \times 100 = (10 - W_d) \times 10$$

CARBON-TO-NITROGEN RATIO

The second important consideration for successful composting is the balance between carbon and nitrogen. High-carbon and high-nitrogen materials should be mixed to achieve a C:N ratio of roughly 30:1 (Table 3–2). High-carbon materials usually are brown or woody. They include autumn leaves, wood chips, sawdust, and shredded paper. High-nitrogen materials generally are characterized as green and include grass clippings, plant cuttings, and fruit and vegetable scraps.

Table 3–2. C:N Ratios of Common Compost Ingredients.[3] *(These data should be viewed as representative ranges, not as universal values.)*

Materials High in Carbon	C:N
autumn leaves	40–80:1
sawdust	200–750:1
wood chips or shavings—hardwood	450–800:1
wood chips or shavings—softwood	200–1,300:1
bark—hardwood	100–400:1
bark—softwood	100–1,200:1
straw	50–150:1
mixed paper	100–200:1
newspaper	400–900:1
corrugated cardboard	600:1
Materials High in Nitrogen	**C:N**
vegetable scraps	10–20:1
fruit wastes	20–50:1
coffee grounds	20:1
grass clippings	10–25:1
cottonseed meal	10:1
dried blood	3:1
horse manure	20–50:1

Although tree leaves in general are considered to be high in carbon and low in nitrogen, this varies considerably according to the species or genus (Table 3–3).

Table 3–3. C:N Ratios of Common Tree Leaves.[4]

Tree	C:N of Leaves	Tree	C:N of Leaves
Alder	15:1	Birch	50:1
Ash	21:1	Aspen	63:1
Elm	28:1	Spruce	48:1
Black elder	22:1	Beech	51:1
Hornbeam	23:1	Red oak	53:1
Linden	37:1	Pine	66:1
Maple	52:1	Douglas fir	77:1
Oak	47:1	Larch	113:1

OTHER CONSIDERATIONS

In addition to moisture content and C:N ratios, several other considerations are important when choosing materials to compost. Prime among these is the time that the materials will need to break down. Woody materials, including wood chips, branches, and twigs, can take up to two years to decompose unless they are finely chipped or shredded. Other materials that break down slowly include corn cobs and stalks, sawdust, straw, and nut shells. When shredded or chipped, these materials can be used as bulking agents to increase aeration within the pile.

The organic materials may be placed in the compost pile all at once (batch composting) or added over time as they become available. If the materials are added gradually, the pile will not heat up significantly until it is large enough to be self-insulating. Although in the past composters often layered various materials, it is probably better to mix the different ingredients to ensure optimal C:N ratios and moisture contents throughout the pile. Always cover food scraps with a layer of sawdust, leaves, or finished compost to trap odors and to avoid attracting flies or other pests.

Research Possibility: *What is the effect of layering versus mixing organic ingredients on the compost pile temperature profile?*

Almost all natural organic materials will compost, but not everything belongs in a classroom or school-yard compost pile. Some organic materials, such as meat and dairy products and oily foods, should be avoided because they tend to attract rodents, raccoons, and other pests. Cat and dog manures may contain harmful pathogens that are not killed during composting. If you plan to use the compost for growing plants, you should avoid adding pernicious weeds because their seeds or runners might survive the composting process. Some weeds to avoid include bindweed, comfrey, and rhizomatous grasses such as Bermuda and crab grass.

CALCULATIONS FOR THERMOPHILIC COMPOSTING

by Tom Richard and Nancy Trautmann

You can use the algebraic equations below to compute the best combination of compost ingredients. The calculations are optional, but they can be useful in a number of situations, such as for figuring out what quantity of wood chips to mix with a known quantity of food scraps from the cafeteria. These calculations also provide an opportunity to apply algebra to a real-world problem.

MOISTURE

The following steps outline how to design your initial mix so that it will have a suitable moisture level for optimal composting.

1. Using the procedure on p. 44, measure the moisture content of each of the materials that you plan to compost. Suppose, for example, that you weigh 10 g of grass clippings (W_w) into a 4-g container and that after drying, the container plus clippings weigh 6.3 g. Subtracting the 4-g container leaves 2.3 g as the dry weight (W_d) of your sample. Using Equation 1, the percent moisture would be:

$$M_n = \frac{W_w - W_d}{W_w} \times 100$$

$$= \frac{10 - 2.3}{10} \times 100$$

= 77% for grass clippings

2. Choose a moisture goal for your compost mixture. Most literature recommends a moisture content of 50–60% by weight for optimal composting conditions. Note that the grass clippings exceed this goal. That is why piles of fresh grass clippings turn into a slimy mess unless they are mixed with a drier material such as leaves or wood chips.

3. Calculate the relative amounts of materials to combine to achieve your moisture goal. The general formula for percent moisture is:

Equation 2:

$$G = \frac{(W_1 \times M_1) + (W_2 \times M_2) + (W_3 \times M_3) + \ldots}{W_1 + W_2 + W_3 + \ldots}$$

in which:

    G = *moisture goal (%)*
    W_n = *mass of material **n** ("as is," or "wet weight")*
    M_n = *moisture content (%) of material **n***

You can use this formula directly to calculate the moisture content of a mixture of materials, and then try different combinations until you get results within a reasonable range. Although this trial and error method will work to determine suitable compost mixtures, there is a faster way. For two materials, the general equation can be simplified and solved for the mass of a second material (W_2) required to balance a given mass of the first material (W_1). Note that the moisture goal must be between the moisture contents of the two materials being mixed.

Equation 3:
$$W_2 = \frac{(W_1 \times G) - (W_1 \times M_1)}{M_2 - G}$$

For example, suppose you wish to compost 10 kg of grass clippings (moisture content = 77%) mixed with leaves (moisture content = 35%). You can use Equation 3 to calculate the mass of leaves needed in order to achieve a moisture goal of 60% for the compost mix:

$$W_2 = \frac{(10 \times 60) - (10 \times 77)}{35 - 60}$$

= 6.8 kg leaves needed

The moisture content and weights for mixtures of three materials can be derived in a similar way, although the algebra is more complicated. For an exact solution, you need to specify the amounts of two of the three materials, and the moisture contents of all three. Then, you can use Equation 4 to determine the desired quantity of the third ingredient.

Equation 4:
$$W_3 = \frac{(G \times W_1) + (G \times W_2) - (M_1 \times W_1) - (M_2 \times W_2)}{M_3 - G}$$

CARBON-TO-NITROGEN RATIO

To calculate the amounts of various materials to add to a compost pile with a C:N ratio of 30:1, you need to know the nitrogen and carbon contents of the individual ingredients. You can use the typical C:N ratios shown in Table 3–3 (p. 45) to calculate the carbon content of an ingredient provided that you know its nitrogen content, or its nitrogen content provided that you know its carbon content. The carbon content of your compost ingredients can be estimated using the **Organic Matter Content** procedure outlined in Chapter 5 (p. 87). Or, you can have your compost ingredients tested for nitrogen and carbon at a soil nutrient laboratory or environmental testing laboratory.

Once you have the percent carbon and nitrogen for the materials you plan to compost, Equation 5 enables you to figure out the C:N ratio for the mixture as a whole:

> **Equation 5:**
> $$R = \frac{W_1[C_1 \times (100-M_1)] + W_2[C_2 \times (100-M_2)] + W_3[C_3 \times (100-M_3)] + \ldots}{W_1[N_1 \times (100-M_1)] + W_2[N_2 \times (100-M_2)] + W_3[N_3 \times (100-M_3)] + \ldots}$$
>
> *in which:*
>
> R = *C:N ratio of compost mixture*
> W_n = *mass of material* n *("as is," or "wet weight")*
> C_n = *carbon (%) of material* n
> N_n = *nitrogen (%) of material* n
> M_n = *moisture content (%) of material* n

This equation can be solved exactly for a mixture of two materials if you know their carbon, nitrogen, and moisture contents. You specify the C:N ratio goal and the mass of one ingredient, then calculate the mass of the second ingredient. Equation 6 is derived by limiting Equation 5 to two ingredients and rearranging the terms to allow you to solve for the mass of the second ingredient:

> **Equation 6:**
> $$W_2 = \frac{W_1 \times N_1 \times \left(R - \frac{C_1}{N_1}\right) \times (100-M_1)}{N_2 \times \left(\frac{C_2}{N_2} - R\right) \times (100-M_2)}$$

Returning to the previous example of grass and leaves, assume the nitrogen content of the grass is 2.4% while that of the leaves is 0.75%, and the carbon contents are 45% and 50% respectively. Simple division shows that the C:N ratio of the grass is 19:1, and the C:N ratio of the leaves is 67:1. For the same 10 kg of grass that we had before, if our goal is a C:N ratio of 30:1, the solution to Equation 6 is:

$$W_2 = \frac{10 \times 2.4 \times (30-19) \times (100-77)}{0.75 \times (67-30) \times (100-35)}$$

= 3.5 kg

Note that we need only 3.5 kg leaves to balance the C:N ratio, compared with 6.8 kg leaves needed to achieve the 60% moisture goal according to our previous moisture calculation. If the leaves were wetter or had a higher C:N ratio, the difference would be even greater. Given the disparity between these results, how should you decide how many leaves to add?

If we solve Equation 5 for the 10 kg of grass and the 6.8 kg of leaves (determined from the moisture calculation) and use the same values for percent moisture, C, and N, the resulting C:N ratio is approximately 37:1. In contrast, if we solve Equation 2 for 10 kg of grass and only 3.5 kg of leaves, we get a moisture content over 66%. (To gain familiarity with using the equations, check these results on your own.)

For mixtures toward the wet end of optimum (>60% moisture content), moisture is the more critical variable. Thus, for the example above, it is best to err on the side of a high C:N ratio. This may slow down the composting process slightly, but it is more likely to avoid anaerobic conditions. If, on the other hand, your mixture is dry, you should optimize the C:N ratio and add water as required.

---

[1] Portions of this chapter were adapted from Dickson, N., T. Richard and R. Kozlowski. 1991. *Composting to Reduce the Waste Stream*. Northeast Regional Agricultural Engineering Service. 152 Riley-Robb Hall, Cornell University, Ithaca NY 14853.

[2] Rynk, R., ed. 1992. *On-Farm Composting Handbook*. Northeast Regional Agricultural Engineering Service, 152 Riley-Robb Hall, Cornell University, Ithaca NY 14853.

[3] Rynk, R., ed. 1992. *On-Farm Composting Handbook*. Northeast Regional Agricultural Engineering Service. This manual includes a table of C:N ratios and %N of a wide range of compost materials. *Available on-line at:*
http://www.cals.cornell.edu/dept/compost/OnFarmHandbook/apa.taba1.html,
or in print from: Northeast Regional Agricultural Engineering Service, 152 Riley-Robb Hall, Cornell University, Ithaca NY 14853.

[4] Schaller, F. 1968. *Soil Animals*. University of Michigan Press, Ann Arbor.

4
MONITORING THE COMPOSTING PROCESS

As decomposition proceeds, a number of changes occur in the physical, chemical, and biological characteristics of the compost mix. Monitoring these changes allows you to assess the progress of your compost, identify potential problems, and compare systems with different initial conditions or ingredients.

Simple observation of the physical changes that occur during composting is one form of monitoring. It is useful to keep a log book, not only to record data but also to note daily observations about the appearance of the compost. Does it appear soggy or dry? Is it shrinking in volume? Is there any odor? Any leachate? At what point do the various types of ingredients become unrecognizable? Have flies or other pests become a problem? If problems do develop during the course of composting, steps can be taken to correct them (Table 4–1).

Another form of monitoring is to take periodic measurements of variables such as the temperature, moisture content, pH, and biological activity. This chapter presents techniques for monitoring these physical, chemical, and biological characteristics of compost. Students can design and conduct a wide array of experiments using these monitoring techniques.

Table 4–1. Troubleshooting Compost Problems.

Symptom	Problem	Solution
Pile is wet and smells like a mixture of rancid butter, vinegar, and rotten eggs	Not enough air	Turn pile
	Or too much nitrogen	Mix in straw, sawdust, or wood chips
	Or too wet	Turn pile and add straw, sawdust, or wood chips; provide drainage
Pile does not heat up	Pile is too small	Make pile larger or provide insulation
	Or pile is too dry	Add water while turning the pile
Pile is damp and sweet smelling but will not heat up	Not enough nitrogen	Mix in grass clippings, food scraps, or other sources of nitrogen
Pile is attracting animals	Meat and other animal products have been included	Keep meat and other animal products out of the pile; enclose pile in 1/4-inch hardware cloth
	Or food scraps are not well covered	Cover all food with brown materials such as leaves, woodchips, or finished compost

TEMPERATURE

Temperature is one of the key indicators of changes occurring during thermophilic composting. If the compost does not heat up, it may be deficient in moisture or nitrogen (Table 4–1). Once the compost does heat up, temperature provides the best indicator of when mixing is desirable (see Figure 1–2, p. 4).

To take temperature readings, use a probe that reaches deep into the compost. Leave the probe in place long enough for the reading to stabilize, then move it to a new location. Take readings in several locations, including various distances from the top and sides. Compost may have hotter and colder pockets depending on spatial variability in the moisture content and chemical composition of the ingredients. Can you find temperature gradients with depth? Where do you find the hottest readings? For systems in which air enters at the bottom, the hottest location tends to be in the core, about two-thirds of the way up. You might expect it to be in the exact center, where insulation by surrounding compost is the greatest, but the core temperatures are affected by the relatively cool air entering at the bottom and warming as it rises through the compost.

Your students might decide to design compost experiments to look for variables or combinations of variables that produce the highest temperatures in the shortest amount of time, or perhaps those variables that maintain hot temperatures for the longest period. One useful way to present your data is to plot the maximum temperature and the time to reach maximum temperature for each compost system as a function of the experimental variable. For example, you could plot the maximum temperature versus the initial moisture content of the compost ingredients. A second graph could show the time to reach maximum temperature versus the initial moisture content.

MOISTURE

Composting proceeds best at moisture contents of 50–60% by weight. During composting, heating and aeration cause moisture loss. That's OK—you want finished compost to be drier than the initial ingredients. Sometimes, however, adding water may be necessary to keep the compost from drying out before decomposition is complete. If the compost appears to be dry, water or leachate can be added during turning or mixing. Below a moisture content of 35–40%, decomposition rates are greatly reduced; below 30% they virtually stop. Too much moisture, on the other hand, is one of the most common factors leading to anaerobic conditions and resulting odor problems.

When you are choosing and mixing your compost ingredients, you may wish to measure the moisture content using the procedure in Chapter 3 (p. 44). After composting is underway, you probably don't need to repeat this measurement because you can observe whether appropriate moisture levels are being maintained. For example, if your compost appears wetter than a wrung-out sponge and starts to smell bad, mix in absorbent material such as dry wood chips, cardboard pieces, or newspaper strips to alleviate the problem. If you are composting in a bioreactor with drainage holes, excess moisture will drain out as leachate. You may find it useful to record the amount of leachate produced by each system, for comparison with initial moisture content, temperature curves, or other variables. If you have a microscope available, try observing a sample of leachate—you will probably find that it is teeming with microbial life.

If you are blowing air through your compost system, you will need to be careful not to create conditions that are too dry for microbial growth. If the temperature drops sooner than expected and the compost feels dry to the touch, moisture may have become the limiting factor. Try mixing in some water and see if the temperature rises again.

ODOR

A well-constructed compost system should not produce offensive odors, although it will not always be odor-free. You can use your nose to detect potential problems as your composting progresses. For example, if you notice an ammonia odor, your mix is probably too rich in nitrogen (the C:N ratio is too low), and you should mix in a carbon source such as leaves or wood shavings.

If compost is too wet or compacted, it will become anaerobic and produce hydrogen sulfide, methane, and other odorous compounds that are hard to ignore. If this occurs in indoor bioreactors, you may wish to take them outside or vent them to the outside, then mix in additional absorbent material such as wood chips or pieces of paper egg cartons. Make sure that you do not pack down the mixture; you want it to remain loose and fluffy to allow air infiltration. In an outdoor compost pile, turning the pile and mixing in additional high-carbon materials such as wood chips should correct the anaerobic condition, although initially the mixing may make the odor even more pronounced.

pH

Why is compost pH worth measuring? Primarily because you can use it to follow the process of decomposition. As composting proceeds, the pH typically drops initially, then rises to 8 or 9 during the thermophilic phase, and then levels off near neutral (see Figure 1–4, p. 8).

At any point during composting, you can measure the pH of the mixture. While doing this, keep in mind that your compost is unlikely to be homogeneous. You may have found that the temperature varied from location to location within your compost, and the pH is likely to vary as well. Therefore, you should plan to take samples from a variety of spots. You can mix these together and do a combined pH test, or you can test each of the samples individually. In either case, make several replicate tests and report all of your answers. (Since pH is measured on a logarithmic scale, it does not make sense mathematically to take a simple average of your replicates. Instead, either report all of your pH values individually, or summarize them in terms of ranges rather than averages.)

pH can be measured using any of the following methods. Whichever method you choose, make sure to measure the pH as soon as possible after sampling so that continuing chemical changes will not affect your results. Also, be consistent in the method that you use when comparing different compost mixtures.

pH PAPER

The least expensive option for measuring compost pH is to use indicator paper. If the compost is moist but not muddy, you can insert a pH indicator strip into the mixture, let it sit for a few minutes to become moist, and then read the pH using color comparison. If the compost is too wet, this technique will not work because the indicator colors will be masked by the color of the mud.

SOIL TEST KIT

Test kits for analysis of soil pH can be used without modification for compost samples. Simply follow the manufacturer's instructions. These kits also rely on color comparison, but the color develops in a compost-water mixture rather than on indicator paper. Soil pH kits are available from garden stores or biological supply catalogs for $5 or more, depending on the number and accuracy range of the tests.

ELECTRONIC METER

The most accurate, but also the most expensive and time consuming method of measuring compost pH, is with a meter. First, you must calibrate the meter by using solutions of known pH. Next, mix the compost with distilled water to make a suspension. Since the amount of water affects the pH reading, it is important to be consistent in the ratio of compost to water and to start with air-dry compost. Finally, place the electrode into a compost/water solution and take a reading. For a detailed description of this method, see *Methods of Soil Analysis*.[1]

MICROORGANISMS[2]

by Elaina Olynciw

A wide range of bacteria and fungi inhabit compost, with species varying over time as changes occur in the temperature and the available food supply. You can simply observe compost microorganisms under a microscope, or culture them for more detailed observation. A third possibility is to measure their metabolic activity, which does not indicate what types or populations of microbes are present but does give an indication of their level of enzymatic activity.

OBSERVING COMPOST MICROORGANISMS

USE: To make simple observations of the microbial communities in compost. Comparisons can be made over the course of several weeks or months as the compost heats up and later returns to ambient temperature.

MATERIALS
- compound microscope
- microscope slides and cover slips
- eye dropper
- 0.85% NaCl (physiological saline)

For bacterial staining (optional)
- 1.6 g methylene blue chloride
- 100 ml 95% ethanol
- 100 ml of 0.01% aqueous solution of KOH
- distilled water
- Bunsen burner
- toothpick
- blotting paper or filter paper

PROCEDURE

1. Make a wet mount by putting a drop of water or physiological saline on a microscope slide and transferring a small amount of compost to the drop. Make sure that you do not add too much compost, or you will have insufficient light to observe the organisms.
2. Stir the compost into the water or saline (the preparation should be watery), and apply a cover slip.
3. Observe under low and high power. You might be able to see many nematodes squirming and thrashing around. Other possibilities include flatworms, rotifers (notice the rotary motion of cilia at the anterior end of the rotifer and the contracting motion of the body), mites, springtails, and fast-moving protozoa. Strands of fungal mycelia may be visible but difficult to recognize (see #5 below). Bacteria appear as very tiny round particles that seem to be vibrating in the background.
4. If you want to highlight the bacteria and observe them in greater detail, you can prepare stained slides:

a) Prepare methylene blue stain by adding 1.6 g methylene blue chloride to 100 ml 95% ethanol, then mixing 30 ml of this solution with 100 ml of 0.01% aqueous solution of KOH.

b) Using a toothpick, mix a small amount of compost with a drop of physiological saline on a slide, and spread it into a thin layer.

c) Let the mixture air-dry until a white film appears on the slide.

d) Fix the bacteria to the slide by passing the slide through a hot flame several times.

e) Stain the slide by flooding it with the methylene blue stain for one minute. Rinse the slide with distilled water and gently blot it dry using blotting or filter paper.

5. Fungi and actinomycetes may be difficult to recognize with the above technique because the entire organism (including the mycelium, reproductive bodies, and individual cells) will probably not remain together. Fungi and actinomycetes will be observed best if you can find them growing on the surface of the compost heap. The growth looks fuzzy or powdery. Lift some compost with the sample on top, and prepare a slide with a cover slip to view under the microscope. You should be able to see the fungi well under 100x and 400x. The actinomycetes can be heat-fixed and Gram-stained to view under oil immersion at 1000x (See procedure under **Culturing Bacteria**, p. 57–59).

6. To collect nematodes, rotifers, and protozoa for observation, use the **Wet Extraction** (p. 69).

CULTURING BACTERIA

USE: To culture bacteria, for colony counts or observation of specific microorganisms.

MATERIALS

Growth media
- 2 g trypticase soy agar (TSA)
- 7.5 g bacto agar
- 500 ml distilled water

For making plates
- 100 ml 0.06M $NaHPO_4/NaH_2PO_4$ buffer (approximately 4:1 dibasic:monobasic, pH 7.6)
- 5 test tubes
- 5 1-ml pipettes
- 5 0.1-ml pipettes
- 3 petri dishes
- glass spreader (glass rod bent like a hockey stick)
- autoclave
- blender
- compost sample, air-dried

For making and observing slides
- inoculating needle
- 0.85% NaCl (physiological saline)
- Bunsen burner
- microscope and slides
- ethanol

For staining slides:
either:
- a Gram stain kit

or:
- 2 g crystal violet
- 100 ml 95% ethanol
- 80 ml 1% ammonium oxalate
- 1 g iodine crystals
- 3 g potassium iodide
- 500 ml distilled water
- amber-colored bottle
- 2.5 g safranin

PROCEDURE

1. Make the TSA media by mixing 2 g TSA, 7.5 g bacto agar, and 500 ml distilled water. Autoclave for 20 minutes, cool until comfortable to touch, then pour into sterile petri dishes and allow to solidify.
2. Autoclave 100 ml 0.06M $NaHPO_4/NaH_2PO_4$ buffer solution in an Erlenmeyer flask. Also autoclave five test tubes, each containing 9 ml of buffer solution, five 1-ml pipettes, and five 0.1-ml pipettes.

3. Autoclave the blender, or sterilize it by rinsing with ethanol. In the blender, mix 5 g compost with 45 ml sterilized buffer solution for 40 seconds at high speed.
4. Perform serial dilutions to 10^{-7}:
    a) Label the five test tubes containing sterilized buffer solution: 10^{-2}, 10^{-3}, 10^{-4}, 10^{-5}, and 10^{-6}.
    b) The mixture in the blender is a 10^{-1} dilution, since 5 g compost are mixed with 45 ml solution (which is equivalent to 45 g). Using a sterile pipette, transfer 1 ml of the compost/buffer mixture into the first test tube, labeled 10^{-2}.
    c) Mix thoroughly, then use another sterile pipette to transfer 1 ml of the solution in the 10^{-2} test tube to the one labeled 10^{-3}. The solution in each test tube will be ten times more dilute than the previous solution from which it was made.
    d) Continue with this sequence until all five test tubes have been inoculated.
    e) Label three petri dishes with the type of compost, date, and dilution: 10^{-5}, 10^{-6}, and 10^{-7}.
    f) Sterilize a glass spreading rod by holding it in the Bunsen burner flame until heated, then cool thoroughly.
    g) Using a sterile pipette, transfer 0.1 ml from the 10^{-4} test tube onto the agar in the petri dish labeled 10^{-5}. Spin the dish and spread the liquid as evenly as possible using the glass spreading rod. Follow this same procedure using 0.1 ml from the 10^{-5} test tube into the dish labeled 10^{-6}, and so forth until all dilutions have been plated.

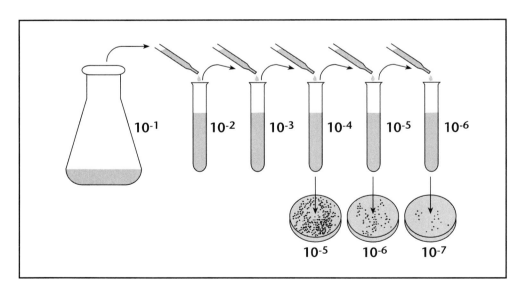

5. Cover the petri dishes and turn them upside down so that any water that condenses will not drip into the cultures. Incubate them at 28°C if possible. (If you do not have access to an incubator, room temperature is acceptable.)

6. After four days, count the colonies and prepare microscope slides. Use an inoculating needle to add a drop of saline to a clean slide. Take a sample of a single bacterial colony and mix it into the saline. Let it air-dry until a white film appears. Heat-fix the slide by passing it through a flame a few times.
7. To highlight bacteria and actinomycetes on the slides, treat them with Gram stain (below).

Gram Staining

1. Prepare Gram stains (unless using a prepared kit):
   Crystal violet:
   Dissolve 2 g crystal violet in 20 ml 95% ethanol. Add this solution to 80 ml of a 1% ammonium oxalate solution. Let the mixture stand for 24 hours, then filter it.
   Gram iodine:
   Add 1 g iodine and 3 g potassium iodide to 300 ml distilled water. Store this solution in an amber bottle.
   Decolorizer: 95% ethanol
   Safranin:
   Add 2.5 g safranin to 10 ml 95% ethanol. Add this solution to 100 ml distilled water.
2. Flood slide with crystal violet for 20 seconds.
3. Gently rinse slide by dipping into a beaker of distilled water.
4. Flood slide with Gram iodine for 1 min. Gently rinse as above.
5. Decolorize by tilting slide and applying 95% ethanol one drop at a time, stopping as soon as no more color is washed out. Gently rinse. At this point, gram-positive bacteria will be purple colored and gram-negative bacteria will be colorless.
6. Flood with safranin for 20 seconds, then gently rinse. This step stains the gram-negative bacteria, and they become pink.
7. Air-dry or gently blot dry. Observe under oil-immersion lens.

CULTURING ACTINOMYCETES

USE: To culture actinomycetes, making it possible to count colonies or to observe individual species.

MATERIALS

Growth media
- 0.4 g typticase soy agar (TSA)
- 10.0 g bacto agar
- 500 ml distilled water
- 10 mg polymixin B in 10 ml 70% ethanol

For making plates
- 100 ml 0.06M $NaHPO_4/NaH_2PO_4$ buffer (approximately 4:1 dibasic:monobasic, pH 7.6)
- 500 ml distilled water
- 5 test tubes
- 5 1-ml pipettes
- 5 0.1-ml pipettes
- 3 petri dishes
- glass spreader (glass rod bent like a hockey stick)
- autoclave
- blender
- compost sample, air-dried

For making and observing slides
- inoculating needle
- scalpel
- 0.85% NaCl (physiological saline)
- Bunsen burner
- microscope and slides

PROCEDURE

1. Mix 0.4 g TSA, 10.0 g bacto agar, and 500 ml distilled water. Autoclave for 20 minutes, then cool to the touch. Add the polymixin B and pour into sterile petri dishes.
2. Autoclave 100 ml 0.06M $NaHPO_4/NaH_2PO_4$ buffer solution.
3. Autoclave the blender, or sterilize it by rinsing with ethanol. In the blender, mix 5 g compost with 45 ml sterilized buffer solution for 40 seconds at high speed.
4. Perform serial dilutions to 10^{-7} using Step 4 of the **Culturing Bacteria** procedure (p. 58).
5. Cover the petri dishes and turn them upside down so that any water that condenses will not drip into the cultures. Incubate them at 28°C if possible. (If you do not have access to an incubator, room temperature is acceptable.)
6. After 14 days, take counts and samples of actinomycetes colonies. Many of the colonies will look powdery white. However, some may take on a rough appearance and produce a variety of pigments. Use an inoculating needle to add a drop of saline to a clean microscope slide. Lifting with a scalpel is a good way to get an intact portion of an

actinomycete colony onto a slide. Allow to air-dry until a white film appears. Heat-fix the slide by passing it through a flame a few times.
7. Observe under a microscope. The actinomycetes probably will be too thick to observe on most of the slide, but at the edges of the colony you will be able to see the pattern that the filaments form. Gram staining can be used to highlight the actinomycetes (see procedure under **Culturing Bacteria**, p. 59).

CULTURING FUNGI

USE: To culture fungi, making it possible to count colonies or to observe individual species.

MATERIALS

Growth media
- 6.5 g potato dextrose agar (PDA)
- 5.0 g bacto agar
- 500 ml distilled water
- 15 mg rifampicin in 10 ml methanol
- 15 mg penicillin G in 10 ml 70% ethanol

For making plates
- 100 ml 0.06M NaHPO4/NaH2PO4 buffer (approximately 4:1 dibasic:monobasic, pH 7.6)
- 3 test tubes
- 3 1-ml pipettes
- 3 0.1-ml pipettes
- 3 petri dishes
- autoclave
- blender
- compost sample, air-dried

For making and observing slides
- scalpel
- microscope, slides, and cover slips

PROCEDURE

1. Mix 6.5 g PDA, 5.0 g bacto agar, and 500 ml distilled water. Autoclave for 20 minutes. Cool until comfortable to handle, then add the antibiotics rifampicin and penicillin G and pour the mixture into sterile petri dishes.
2. Autoclave 100 ml 0.06M $NaHPO_4/NaH_2PO_4$ buffer solution in an Erlenmeyer flask. Also autoclave three test tubes, each containing 9 ml of buffer solution, three 1-ml pipettes, and three 0.1-ml pipettes.
3. Autoclave the blender, or sterilize it by rinsing with ethanol. In the blender, mix 5 g compost with 45 ml sterilized buffer solution for 40 seconds at high speed.
4. Perform serial dilutions to 10^{-5}:
   a) Label the three test tubes containing buffer solution: 10^{-2}, 10^{-3}, and 10^{-4}.
   b) The mixture in the blender is a 10^{-1} dilution, since 5 g compost are mixed with 45 ml solution (which is equivalent to 45 g). Using a sterile pipette, transfer 1 ml of the compost/buffer mixture into the first test tube, labeled 10^{-2}.
   c) Mix thoroughly, then use another sterile pipette to transfer 1 ml of the solution from the 10^{-2} test tube to the one labeled 10^{-3}. The solution in each test tube will be ten times more dilute than the previous solution from which it was made.

d) Continue with this sequence until all three test tubes have been inoculated.
e) Using a sterile pipette, transfer 0.1 ml from the 10^{-2} test tube onto the agar in the petri dish labeled 10^{-3}. Spin the dish and spread the liquid as evenly as possible using the glass spreading rod. Follow this same procedure using 0.1 ml from the 10^{-3} test tube into the dish labeled 10^{-4} and so forth until all dilutions have been plated.

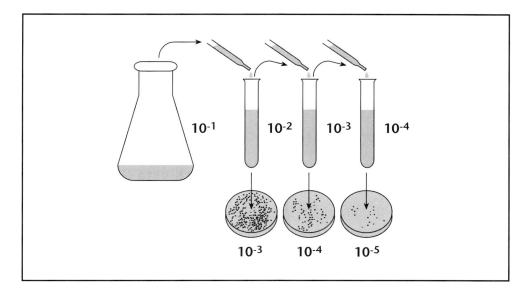

5. Cover the petri dishes and turn them upside down so that any water that condenses will not drip into the cultures. Incubate them at 28°C if possible. (If you do not have access to an incubator, room temperature is acceptable.)
6. After three days, take counts and samples of fungal colonies. Lift a portion of a fungal colony intact onto a clean slide (it will still be attached to the agar), add a cover slip, and observe without staining. Look at the edges of the colony where the sample will be thin enough for light to pass through.

MEASURING MICROBIAL ACTIVITY
FLUORESCEIN DIACETATE (FDA) HYDROLYSIS AS A MEASURE OF TOTAL MICROBIAL ACTIVITY IN COMPOSTS AND SOILS[3]

USE: To measure the level of metabolic activity of microbes in a compost sample.

BACKGROUND
During the course of composting, the size and activity level of microbial populations will vary. For example, as the temperature drops at the end of the thermophilic phase, activity levels may be relatively low because the populations of heat-loving microbes are diminishing and the mesophilic microorganisms are just beginning to recolonize. Another time of low activity is at the end of composting, after the curing process has left little available carbon to support microbial growth.

The FDA test measures microbial activity through color change. As microorganisms hydrolyze fluorescein diacetate to fluorescein, the solution changes from colorless to yellow. The amount of yellow pigment measured by a spectrophotometer will give an indication of how much hydrolysis has occurred, and, therefore, it will give an indication of the total metabolic activity of the microbial population.

MATERIALS
- 20 mg fluorescein diacetate (FDA)
- 10 ml acetone ACS grade
- 20 ml 0.06M sodium phosphate buffer solution (pH 7.6)
- 14 125-ml Erlenmeyer flasks
- 14 small test tubes with screw caps
- 4-cm Büchner funnel
- Whatman #1 filter paper
- 125-ml side flask for vacuum filter
- shaker (If there are enough students in a group doing this experiment, they can take turns swirling the flasks for a total of 20 minutes.)
- centrifuge
- small test tubes (to fit in centrifuge)
- spectrophotometer
- drying oven
- 10-g sample of fresh compost

METHOD
1. Place 0.5 g compost in a 105°C oven overnight for dry weight determination.
2. Make duplicate samples by placing undried compost equivalent to 0.5 g dry weight in each of two 125-ml flasks.
3. Add 20 ml of phosphate buffer solution to each flask.
4. Make FDA stock solution by mixing 20 mg FDA in 10 ml acetone. Add 0.2 ml of FDA stock solution to each flask containing compost and buffer solution.
5. Place flasks on a shaker (90 rpm) at 25°C for 20 minutes.

6. Using a Büchner funnel, filter through Whatman #1 paper with a light vacuum.
7. Transfer 2 ml of the filtrate into a small test tube and add 2 ml acetone to stop the reaction.
8. Centrifuge samples for 5 minutes at 5000 rpm to remove particulates.
9. Using a spectrophotometer, read the absorbance at 500 nm against a buffer blank.

Preparation of a standard curve

1. Make duplicate samples of a range of FDA concentrations by adding 0, 0.02, 0.04, 0.05, 0.08, and 0.1 ml FDA stock solution to 5 ml of PO_4 buffer solution in each of 10 capped test tubes.
2. Cap tightly and heat in a boiling water bath for 60 minutes to hydrolyze the FDA.
3. Cool 10 minutes and add hydrolyzed FDA to flasks containing 0.5 g compost and 15 ml phosphate buffer. (Note: You will have to set up a different standard curve for each compost you are analyzing because a certain percentage of the fluorescein will adhere to the compost and this will vary from compost to compost.)
4. Place on a shaker or swirl fairly rapidly and consistently for 20 minutes.
5. Follow Steps 6-9 in the Methods section above.
6. Record your absorbance data using the table below.

FDA solution (ml)	FDA (ug)	ABSORBANCE		
		Rep. #1	Rep. #2	Mean
0	0			
0.02	40			
0.04	80			
0.05	100			
0.08	160			
0.10	200			

ANALYSIS

Make a standard curve by graphing FDA concentrations versus optical density readings. Then, use the graph to determine the FDA concentrations for your compost samples based on their optical density readings.

By comparing samples from different stages in the composting process, what observations can you make about the relative activity of the microbial populations?

INVERTEBRATES

In outdoor compost piles, a wide range of invertebrates takes part in the decomposition of organic matter. Try monitoring invertebrate life in the pile over the course of the compost process. Do invertebrates appear while the pile is hot, or not until it cools? Do you find different types of organisms during the various stages of decomposition?

In indoor composting you may find fewer (or no) invertebrates. In vermicomposting, worms will likely be the only macro-invertebrates unless leaves or dirt were introduced into the bin for bedding. Worms can be observed by using the methods for larger invertebrates that are described in this section.

Three methods are described here for collection of organisms living in compost. For the larger compost invertebrates, such as earthworms, sowbugs, or centipedes, the **Pick and Sort** method works well. To find smaller invertebrates, you may wish to try a **Berlese Funnel**, which concentrates creatures by collecting them in a vial placed below a funnel containing compost. Tiny invertebrates that live in the films of water surrounding the organic matter or soil particles, including nematodes and flatworms, are best collected by using the **Wet Extraction**, a variation of the Berlese Funnel in which the compost is soaked in water rather than dried.

PICK AND SORT

USE: To collect organisms that are easily visible to the naked eye.

MATERIALS
- light-colored trays or pans
- tweezers or spoons
- jars for temporary sorting and display of organisms
- flashlights (optional)
- magnifying lenses or dissecting microscopes
- petri dish or watch glass for use with microscope
- fresh compost sample, preferably from an outdoor compost pile

PROCEDURE

1. Take samples of compost from various locations in the heap. Are there some organisms that you find near the surface and others only at greater depths? Spread each compost sample in a large tray or pan, or on a large piece of paper, preferably light in color for maximum contrast. When sorting through the compost, students should use soft tweezers, plastic spoons, or other instruments that will not hurt the organisms.

2. Flashlights and magnifying lenses can be used to enhance the observation. The larger organisms, such as worms, centipedes, millipedes, and sowbugs, can easily be seen with the naked eye, but they can be observed more closely under the microscope. Place samples of the compost in petri dishes or watch glasses and observe them under a dissecting microscope.

3. In a classroom setting, you can use an overhead projector to show the outline of the organisms. It can be fun for the students to see the organisms projected to large sizes, crawling around larger than life on the screen. Caution: If you are using live organisms, it is best to have someone standing next to the projector to catch the organisms as they crawl away from the projection platform.

BERLESE FUNNEL

USE: To concentrate into a vial small organisms not easily collected through picking and sorting.

MATERIALS
- ring stand and ring
- funnel lined with small piece of window screen, or kitchen sieve enclosed in paper funnel
- beaker or jar
- light source (25 watt)
- fresh compost sample, preferably from an outdoor compost pile
- magnifying glass or dissecting microscope
- petri dish or watch glass
- optional: 100 ml of 90% ethanol/10% glycerol solution (if you wish to preserve invertebrates)

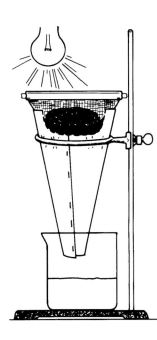

PROCEDURE
1. Assemble a funnel and sieve, either by lining the bottom of a funnel with a small piece of window screen, or by making a funnel to fit over a kitchen sieve.
2. Into a beaker or jar, add 100 ml of water if you wish to collect live organisms or a mixture of 90% ethanol and 10% glycerol if you wish to preserve them. Place the beaker just below the funnel to collect the specimens.
3. Position a light source (25 watt) 2–5 cm above the funnel, or place the collecting apparatus in a sunny location. The light, heat, and drying will gradually drive the compost organisms downward through the funnel and into the collecting jar. If you use too strong a light source, the organisms will dry up and die before making it through the compost and into the funnel.
4. Place compost in the sieve or funnel, making a layer several centimeters deep. Leave for several hours or overnight.
5. Place your collected organisms in a petri dish or watch glass, and observe them under a dissecting microscope or with a magnifying glass.

WET EXTRACTION

USE: To collect nematodes, rotifers, enchytraeids, and other small organisms that live in aqueous films surrounding compost particles.

MATERIALS
- ring stand and attachments
- funnel
- rubber tubing to fit small end of funnel
- pinch clamp
- beaker or jar
- 20-cm² square of cheesecloth
- 25-cm length of string
- fresh compost sample
- light source (25 watt)
- light microscope
- microscope slides and cover slips

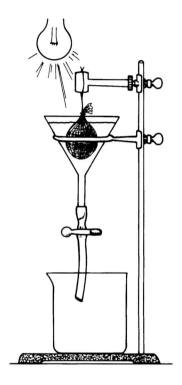

PROCEDURE
1. Assemble the apparatus as shown in illustration, with the funnel suspended above the beaker and the rubber tubing leading from the bottom of the funnel into the beaker. Close the tubing with a clamp.
2. Make a bag of compost by placing a sample on the cheesecloth, gathering the edges, and tying them tightly together at the top. The amount of compost you use will depend on the size of your funnel (the finished bag should be small enough to fit within the funnel with space for water to flow around the edges).
3. Suspend the bag of compost, locating it so that it hangs inside the funnel with clearance around its edges.
4. Fill the funnel with water, making sure that the compost bag is submerged but not sitting on the funnel walls.
5. Place the light above the funnel and turn it on.
6. After 24 hours, open the clamp and allow the water to drain into the beaker.
7. Observe drops of water from the collected sample under the microscope. What types of organism can you identify? You may wish to count organisms from a specified amount of compost, comparing quantities and types of organisms found to those in other types of compost or soil.

[1] Page, A. L., R. H. Miller, and D. R. Keeney, eds. 1982. *Methods of Soil Analysis*, Part 2, 2nd ed. American Society of Agronomy, Inc., Soil Science Society of America, Inc., Madison, WI. pp. 208–209.

[2] The activities in the Microorganisms section were written by Elaina Olynciw, biology teacher at A. Philip Randolph High School in New York City. The descriptions assume a working knowledge of microbiological techniques. For those who have not previously worked with microorganisms, the following laboratory manual provides a good overview of techniques: Selley, H. W., Jr., P. J. Vandemark, and J. J. Lee. 1991. *Microbes in Action*, 4th ed. W. H. Freeman & Co., NY.

[3] Adapted from Schnurer, R. and T. Rosswall. 1982. Fluorescein diacetate hydrolysis as a measure of total microbial activity in soil and litter. *Applied and Environmental Microbiology* 43:1256–1261, and
Craft, C. M. and E. B. Nelsen. 1996. Microbial properties of composts that suppress damping-off and root rot of creeping bentgrass caused by *Pythium graminicola*. *Applied and Environmental Microbiology* 62:1550–1557.

5
COMPOST PROPERTIES

The properties of compost vary widely, depending on the initial ingredients, the process used, and the age of the compost. What properties are important and how can we measure them? Because compost is used primarily in horticulture and agriculture, properties that affect soils and plant growth are important (Table 5–1). The stability and quality tests outlined in this chapter can be used to tell whether a compost is finished and ready to use with plants. We have also included a series of tests that measure how compost affects specific properties of the soil with which it is mixed. Although not covered here, you may want to determine the nutrient status of your compost or compost/soil mix by using standard soil-nutrient test kits available from garden stores or science supply catalogs.

Table 5–1. Tests of Compost Properties.

Type of Test	Name	Properties Tested	Page #
Stability	Jar Test	odor development	73
	Self-Heating Test	heat production	74
	Respiration Test	CO_2 generation	75
Quality	Phytotoxicity Bioassay	effects on seed germination and root growth	79
Effects on soil properties	Porosity	volume of pore space	83
	Water Holding Capacity	ability to retain moisture	85
	Organic Matter Content	percent organic matter	87
	Buffering Capacity	ability to resist change in pH	89

One of the questions that arises in composting is how to tell when the process is finished and the compost is ready for use with plants. Thermophilic composting has two end points, one at the end of the rapid decomposition phase, after which the compost is called "stable," and the second after a several-month period of slower chemical change called "curing." It is after this curing stage that compost is "mature" and ready for use as a soil amendment to enhance plant growth.

To tell if your compost is mature, follow these steps:
1. Monitor the temperature changes. When compost cools and does not reheat after mixing, the period of rapid decomposition has ended.

2. Observe the appearance. Once compost cools, it has probably shrunk to one-half or less of its original volume. It should look brown and crumbly, and it should have a pleasant earthy odor and no recognizable chunks of the initial ingredients. (Wood chips might remain because they are quite slow to break down. They can be screened out for reuse in another batch of compost.) However, if less resistant ingredients such as leaves or banana peels have not decomposed, the process has slowed down because of some constraint other than available food. The moisture level might have become too low, or perhaps the system was too small for adequate heat retention. You may want to correct any problems by using the troubleshooting guide (Table 4–1, p. 51) before performing tests of the various compost properties.
3. Test the stability. If the compost appears to be fully decomposed, you may wish to test whether it is stable, meaning that the phase of rapid decomposition has been completed and the organic materials are no longer rapidly changing. The **Jar Test**, **Self Heating Test**, and **Respiration Test** provide three different ways to assess compost stability. (If used while unstable, compost can impair rather than enhance plant growth because the continuing decomposition uses nitrate and oxygen needed by plants.)

Research Possibility: *These tests for compost stability were developed for thermophilic composting. Can you design a research project to determine whether these tests are useful for vermicompost?*

4. Assess the quality. Once compost is stable, it is not necessarily ready to use with plants. A several-month period of curing allows ammonia, acetic acid, and other intermediate products of decomposition to transform into compounds that will not suppress seed germination, injure plant roots, or stunt plant growth. There is no definitive end point, and the degree of curing needed depends on how the compost will be used. The **Phytotoxicity Bioassay** provides a means of assessing whether the compost contains any substances that are likely to be detrimental to plant growth.

If the compost appears suitable for use with plants, you may want to test its effects on soil properties, including **Porosity**, **Water Holding Capacity**, **Organic Matter Content**, and **Buffering Capacity**, following the procedures outlined in this chapter. The ultimate test of the quality of a compost is its effect on plant growth (see Chapter 6).

Research Possibility: *How does age of compost affect various compost properties (e.g., phytotoxicity, water holding capacity)? How does the initial mix of ingredients affect compost properties?*

Research Possibility: *What is the relationship of various compost properties to growth of specific plants?*

COMPOST STABILITY

JAR TEST

USE: To determine if organic matter is thoroughly decomposed, based on odor development in an enclosed sample.

BACKGROUND

If compost is not yet fully decomposed, it will smell rotten after being moistened and enclosed for a few days. This is because anaerobic conditions develop and noxious compounds such as methane or organic acids are formed. In contrast, stable compost has an insufficient supply of readily degradable organic matter for significant odors to develop.

MATERIALS

- compost sample
- water
- jar or plastic bag that can be tightly sealed

PROCEDURE

Add enough water to a compost sample so that it feels moist but not soggy. Place it in a jar or plastic bag, seal the container, then let it sit for a week at room temperature (20–30°C).

ANALYSIS

When you open the jar or bag at the end of the week, you will be greeted by a pleasant earthy odor if the compost is mature. If it is immature, the smell will be putrid because continued decomposition has depleted the oxygen and caused anaerobic conditions to develop.

Another sign of instability is any visible growth of mold or other fungus. If the organic matter is fully broken down into humus, it will not look fuzzy or slimy after being enclosed for a week.

SELF-HEATING TEST

USE: To determine if organic matter is thoroughly decomposed, based on heat production by microorganisms under optimal conditions.

BACKGROUND

In compost that contains readily degradable organic matter and sufficient moisture, microbial populations will grow rapidly, and their metabolic heat will cause the temperature to rise. If the compost heats up, this is an indication that the organic matter is not yet fully decomposed.

MATERIALS
- compost sample
- gallon-size jar or thermos
- thermometer with 15-cm or longer probe

PROCEDURE
1. Fill a gallon-size container with compost at 40–50% moisture content (see p. 44 for moisture measurement). If your compost is too dry, add distilled water to achieve 50% moisture. If it is too wet, spread the compost in a thin layer to air dry.
2. Seal the container and insulate it with a layer of foam or other insulating material.

ANALYSIS

After two or three days, open the container and measure the compost temperature. If it is more than a few degrees above the ambient air temperature, the compost is not yet stable, meaning that the available organic matter is not yet fully decomposed.

Lack of heating is a more ambiguous result; it does not necessarily indicate that the compost is stable. Perhaps microbial growth was inhibited by lack of nitrogen rather than because the phase of rapid decomposition was complete. See the steps listed on pp. 71–72 to more fully diagnose the meaning of your results.

RESPIRATION TEST[1]

USE: To determine if organic matter is thoroughly decomposed, based on CO_2 production.

BACKGROUND

The CO_2 curve during thermophilic composting looks similar to the temperature curve (p. 2). This makes sense, since both heat and CO_2 are released by microbes as they decompose organic matter. The highest rates of CO_2 production occur during the thermophilic phase, when decomposition rates are at their peak. As the quantities of readily degradable organic matter diminish, the rate of CO_2 production also drops. The Respiration Test provides a measure of whether the rate of CO_2 production has dropped low enough for the compost to be considered stable. It works by capturing the CO_2 gas, which reacts with the NaOH in solution to produce carbonic acid, as shown in the following equation:

$$2\,NaOH + CO_2\text{ (gas)} \rightarrow 2H^+ + CO_3^{2-} + 2Na^+ + O^{2-}$$

The goal of the Respiration Test is to determine whether the readily degradable organic matter has been depleted, causing microbial respiration rates to be low. However, you might find low CO_2 production rates even in an immature compost if microbial growth has been inhibited by unfavorable moisture, pH, or oxygen levels during the compost process. You can avoid this type of false result by also testing your compost with the **Jar Test** (p. 73).

MATERIALS

For the incubation:
- balance
- compost sample
- 2 1-gallon jars (plastic or glass), with lids that form tight seals
- 8 100-ml beakers or jars to hold NaOH (each needs to fit inside a gallon jar)
- tall thin jar, such as a jelly jar, to hold compost sample (needs to fit inside gallon jar, alongside one of the NaOH jars)
- 250 ml 1M NaOH
- 10-ml pipette
- incubator (optional)
- You may find it useful to run the respiration test using potting soil or well-aged compost for comparison with your own compost. In this case, you will need an additional gallon jar with a lid, 4 additional NaOH jars, and one more tall thin jar.

For the titrations:
- 300 ml 1M HCl
- phenolphthalein
- burette
- magnetic stirring plate and bar (optional)

PROCEDURE

This procedure takes about 1 1/2 weeks.

Wednesday to Friday: Standardize compost moisture content

Measure the moisture content of your compost following the procedure on p. 44.

Adjust the moisture level of your compost to 50%. This step is important because moisture content will affect the respiration rate and the stability rating. If the sample that you measured was drier than 50%, add water (to your fresh compost, not to the oven-dried sample), stirring in enough distilled water to bring the moisture level up to 50%.

If the measured sample was wetter than 50%, spread fresh compost into a thin layer to air-dry until the desired moisture level is achieved. Take care not to over-dry the compost because this will decrease its microbial activity.

Friday: Assemble the materials

Assemble the materials needed for the incubation vessels and the titrations. Using a 10-ml pipette, transfer 20 ml of 1M NaOH solution into each of eight 100-ml beakers or jars. Tightly seal the jars.

Weekend: Allow sample to equilibrate

The equilibration step is optional if your compost was fresh and close to the 50% moisture level before adjustment. However, if it was frozen or dried rather than fresh, it will need time for the microbes to adjust to the new conditions. After adjusting the moisture content, put the compost into a jar that has ample air space. Close the lid to preserve moisture, and let it sit over the weekend at room temperature in order to equilibrate.

Monday: Begin incubation

1. Stir compost thoroughly, then transfer 25 g into the tall sample jar.
2. In a 1-gallon jar, place next to each other the compost sample and a jar containing NaOH.
3. In a second 1-gallon jar, create a blank by using a jar of NaOH but omitting the jar of compost.
4. Tightly close the lids of the gallon containers. Record the date, time, and air temperature.
5. Store at room temperature (20–30°C), or warmer if possible. You want to provide a constant warm temperature. A sunny windowsill probably is not appropriate because it will get hot during the day but cold at night. An incubator set at 37°C is ideal. One possibility is to create your own incubator using a light or heating pad in a box.
6. Over the next four days, you will measure the amount of CO_2 absorbed by each NaOH trap. This is accomplished by titrating with 1M HCl according to the procedure outlined below.

Tuesday through Friday: Titration

Each day, open the incubation vessel containing compost and remove the jar of NaOH. Add a fresh jar of NaOH and reseal the incubation vessel. At this point, you can either carry out the titration immediately, or you can save it until Friday if you prefer to carry out all the titrations at once. In this case, tightly seal each jar of NaOH and label it with the date and type of sample. Follow the same procedure for the blank jars containing no compost.

For the titration:

1. Add two to three drops of phenolphthalein indicator to the NaOH solution.
2. Fill the burette with HCl, and zero it. Titrate with acid until the NaOH solution begins to become clear. Agitate by hand or use the magnetic stirrer to mix the solution while adding acid.
3. As the end point gets closer, add acid, one drop at a time, mixing thoroughly between drops. The end point has been reached when the solution turns from pink to clear.

    The greater the amount of CO_2 that has been released from the compost sample and absorbed into the solution, the less acid it will take to reach the titration endpoint. This is because as CO_2 is absorbed, the solution becomes increasingly acidic with the formation of carbonic acid (see equation on p. 75).
4. Record the date and time, the molarity of HCl used, and the volume of HCl required to reach the end point.
5. Friday: Clean out the incubation vessel and calculate your results.

ANALYSIS

Calculate the mass of CO_2 generated by your compost sample:

$$CO_2 \cdot C\ (mg) = \frac{HCl_b - HCl_s}{1000\ ml/l} \times HCl\ molarity\ (mol/l) \times 12g\ C/mol \times 1000\ mg/g$$

where:

HCl_b = *ml HCl used in titration of blank*
HCl_s = *ml HCl used in titration of sample (from jar containing compost)*
$CO_2 \cdot C$ = *mass of CO_2-carbon generated (mg)*

which simplifies to:

$$CO_2 \cdot C\ (mg) = (HCl_b - HCl_s) \times 12$$

Plot CO_2 production over the course of the four days of readings. Compare your data to a baseline obtained by running the respiration test on potting soil or fully decomposed, well-aged compost. This baseline should be a fairly level line close to zero because the microbial activity is minimal in a fully decomposed sample. Unfinished compost should support more microbial growth, so CO_2 production would be expected to be higher.

If you have used an incubator, you can compare the peak respiration rate of your sample to a stability index (Table 5–2). To do this, you will need to express your results in terms of the organic carbon content of the sample. Carrying out the **Organic Matter Content** procedure on p. 87

will give you the percentage of carbon, which you can use to calculate the mass of organic carbon in your sample:

organic carbon (g) = (wet weight of sample)(100 − % moisture)(% carbon)
$$= 25 \text{ g} \times 50\% \times \%C$$
$$= 12.5 \times \%C$$

To standardize your respiration data, divide the $CO_2 \cdot C$ values by the mass of organic carbon in the compost sample. These standardized respiration rates can then be evaluated using the compost stability ratings in Table 5–2.

mg $CO_2 \cdot C$/g organic carbon/day = mass $CO_2 \cdot C$ (mg/day)/organic carbon (g)

Table 5–2. Compost Stability Index.*

Respiration Rate (mg $CO_2 \cdot C$/ g organic carbon/day)	Rating	Trends		
<2	Very stable	Potential for odor generation ↓	Potential for inhibition of plant growth ↓	Potential for inhibition of seed germination ↓
2–5	Stable			
5–10	Moderately stable			
10–20	Unstable			
>20	Extremely unstable			

*The values in this table are based on incubation at 37°C. If your incubation is carried out at a lower temperature, data interpretation using this table may be misleading.

COMPOST QUALITY

PHYTOTOXICITY BIOASSAY[2]

USE: To determine whether a compost contains substances that inhibit seed germination or growth of the radicle (the embryo root).

BACKGROUND

Immature compost may contain substances such as methane, ammonia, or acetic acid that are detrimental to plant growth. These are created during composting and later broken down during the curing phase. Even mature compost may contain substances that inhibit plant growth, such as heavy metals, salts, pesticide residues, or other toxic compounds contained in the original compost ingredients.

One way of testing compost quality is to analyze it chemically. The trouble with this approach is that it is not feasible to test for every compound that might possibly be present. Bioassays, in which test organisms are grown in a water extract of compost, provide a means of measuring the combined toxicity of whatever contaminants may be present. However, they will not identify what specific contaminants are causing the observed toxicity.

To provide a useful measure of toxicity, a bioassay must respond predictably to a range of concentrations of a known compound, as well as to complex mixtures of contaminants. It should also be sensitive, rapid, and cost-effective. Garden cress (*Lepidium sativum*, L.) is commonly used for compost bioassays because it meets these criteria.

MATERIALS
- compost sample (roughly 200 g)
- small pan (5–10 cm for drying compost)
- balance
- drying oven (105°C)
- funnel
- ring stand with attachment to hold funnel
- double layer of cheesecloth, large enough to line funnel
- 100-ml graduated cylinder
- 1,000-ml beaker or jar
- 200-ml beaker or jar
- 15 9-cm petri dishes
- 15 7.5-cm paper filter disks
- tweezers
- metric ruler or caliper
- cress seeds (*Lepidium sativum*, L.)
- 1 liter distilled water
- litmus paper or pH test kit for water

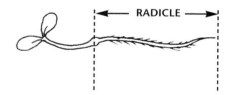

PROCEDURE

Prepare a compost extraction:

1. In order to standardize the dilution from one compost sample to another, you need to correct for the water content of the compost. To do this, first measure the percent moisture of a compost sample (p. 44).

2. The next step is to calculate how much of your wet compost would be equivalent to 100 g dry weight:

$$\_\_\text{g wet compost} = \frac{100 \text{ g dry compost}}{(W_w - W_d)/W_w}$$

3. Moisture content varies from one compost to another, and this needs to be taken into account when determining how much additional water to use for the extraction:

   __ g (or ml) distilled water = 850 g total − __ g wet compost
   to be added for extraction   *from Step 2*

   Add the amount of distilled water calculated in the above equation to the amount of wet compost calculated in Step 2. Stir well, then allow the compost to settle for approximately 20 minutes.

4. Skim off the top 200 ml, and filter it through a double layer of cheesecloth. The filtrate is your extract.

5. Measure and record the pH of the distilled water. If it is not near neutral, either find a new supply or add a small amount of baking soda to buffer the solution, then remeasure the pH.

6. Make a 10x dilution by mixing 10 ml of extract with 90 ml of distilled water.

7. Measure and record the pH of the compost extract and the 10x dilution.

8. In each of 15 9-cm petri dishes, place a 7.5-cm paper filter. Label five dishes "control," another five "10x dilution," and the remaining five "full strength." In addition, you may wish to include information such as the type of compost and the date.

9. To each petri dish, add 1 ml of the appropriate test solution: distilled water, diluted extract, or extract at full strength. Evenly space eight cress seeds in each dish, then cover.

10. Enclose the petri dishes in sealed plastic bags for moisture retention. Incubate for 24 hours in the dark at a steady warm temperature—27°C is ideal. (If you can't maintain this warm a temperature, you may need to lengthen the incubation time.)

11. Open each dish and count how many seeds have germinated. Of these, measure the length of the radicle, the part that looks like a root. Fill in Table 5–3.

Table 5–3. Germination and Radicle Length in Compost Extract.

Treatment	# Germinated	Mean # Germinated	Radicle Length (mm)								Mean Radicle Length* (mm)
Distilled water											
dish #1											
dish #2											
dish #3											
dish #4											
dish #5											
Filtrate (10x dilution)											
dish #1											
dish #2											
dish #3											
dish #4											
dish #5											
Filtrate (full strength)											
dish #1											
dish #2											
dish #3											
dish #4											
dish #5											

* In calculating mean radicle lengths, include only seeds that germinated.

ANALYSIS

1. For each treatment, calculate the percent germination:

$$\% G = \frac{G_t}{G_c} \times 100$$

in which:
%G = *percent germination*
G_t = *mean germination for treatment*
G_c = *mean germination for distilled water control*

2. Calculate the percent radicle length for each treatment:

$$\%L = \frac{L_t}{L_c} \times 100$$

in which:

$\%L$ = percent radicle length
L_t = mean radicle length for treatment
L_c = mean radicle length for distilled water control

3. For each treatment, calculate the germination index (**GI**) and compare it with the ratings in Table 5–4:

$$GI = \frac{\%G \times \%L}{10,000}$$

Table 5–4. Garden Cress Germination Index.*

Germination Index	Rating
1.0–0.8	No inhibition of plant growth
0.8–0.6	Mild inhibition
0.6–0.4	Strong inhibition
<0.4	Severe inhibition

* It is possible for the germination index to be greater than one, in which case the extract enhanced rather than impaired germination and/or radicle growth.

Consider the following questions:
- How does radicle length compare with germination? (Does one indicator of phytotoxicity appear more sensitive than the other?) Why do you think a combination of the two measurements is used?
- What can you conclude about your compost? Is it suitable for use on sensitive plants?
- What do you think the results would be if you tried this test at various stages during the composting process?

Research possibility: *What types of chemical conditions inhibit cress seed germination or radicle growth? Is pH important? Can you identify certain types of compounds, (e.g., salts, nutrients) or conditions (e.g., pH, age of compost) that cause inhibition?*

Research possibility: *Some leaves, such as those of black walnut or eucalyptus trees, contain chemicals that inhibit growth of other plants. Are these compounds broken down by composting?*

EFFECTS ON SOIL PROPERTIES

Compost is referred to as a soil amendment or a substance that is added to soil to improve its physical or chemical characteristics. Many claims are made about how compost enhances soil drainage as well as the ability of soil to hold water so that it is available to microorganisms and plant roots. Are these claims valid? Does compost make clay soils less compact and better drained? Do compost amendments make sandy soils better able to hold water? The tests for **Porosity** and **Water Holding Capacity** will help you to answer questions such as these.

The procedure for measuring **Organic Matter Content** provides a means of comparing different types of compost and soil, or compost in various states of decomposition. Most soils contain less than 20% organic matter, whereas the percentage in compost is much higher. The final test, **Buffering Capacity**, provides a simple measure of one way in which compost can influence the chemistry of soil and of the water that percolates through it.

POROSITY

USE: To measure the volume of pore space in a compost or soil sample.

BACKGROUND

Porosity measures the proportion of a given volume of soil occupied by pores containing air and water. It provides an indication of whether the soil is loose or compacted, which affects both drainage and aeration.

A sandy soil has large particles and large pore spaces whereas a clayey or silty soil has smaller pore spaces. What may be surprising, though, is that the numerous small pores in the clayey or silty soil add up to a larger total pore volume than in a sandy soil. In general, addition of organic matter such as compost increases a soil's porosity.

MATERIALS
- petri dish
- balance with g accuracy
- 100-ml graduated cylinder
- stirring rod slightly longer than graduated cylinder
- sample of compost, soil, or compost/soil mixture

PROCEDURE
1. Fill the graduated cylinder about half full with compost, soil, or compost/soil mixture.
2. Tap the cylinder firmly against your hand several times to settle the sample, then record the volume after it has settled.
3. Pour the sample out and save it to use in Step 5.
4. Fill the graduated cylinder to the 70-ml level with water.
5. Slowly add the compost, soil, or compost/soil mixture saved from Step 3.
6. Stir with rod to break up clumps, then let stand for 5 minutes to allow bubbles to escape.
7. Record the final volume of the compost/water mixture.

ANALYSIS

1. Calculate the volume of solids in your compost or soil:

   vol. of solids (ml) = vol. of compost/water mix (ml) − 70 ml water
   from Step 7

2. Calculate the total pore space volume:

   vol. of pore space (ml) = vol. of packed soil (ml) − vol. of solids (ml)
   from equation above

3. Determine the porosity:

   $$\textbf{porosity} = \textbf{\% pore space} = \frac{\textbf{vol. of pore space}}{\textbf{vol. of packed soil}} \times \textbf{100}$$

WATER HOLDING CAPACITY

USE: To determine the ability of a soil or compost to retain moisture against drainage due to the force of gravity.

BACKGROUND

The water holding capacity of a soil determines its ability to sustain plant life during dry periods. Water is held in the pores between soil particles and in the thin films surrounding particles. Different types of soil retain different amounts of water, depending on the particle size and the amount of organic matter. Organic matter adds to a soil's water holding capacity because humus particles absorb water.

MATERIALS

- funnel
- tubing to attach to bottom of funnel
- clamp for tubing
- ring stand with attachment to hold funnel
- circular filter paper or coffee filter large enough to line funnel
- 100 ml of air-dried compost, soil, or compost/soil mixture
- balance with g accuracy
- 2 250-ml beakers
- 100-ml graduated cylinder
- stirring rod slightly longer than graduated cylinder

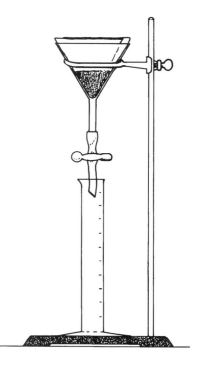

PROCEDURE

1. Spread out and thoroughly air-dry the compost, compost/soil mixture, or soil samples.
2. Attach tubing to the bottom of the funnel and clamp it shut. Attach the funnel to the ring stand, suspended above the graduated cylinder.
3. Line the funnel with filter paper or a coffee filter.
4. Place 100 ml of air-dried compost or compost/soil mixture into the funnel.
5. Using the graduated cylinder, measure 100 ml of water. Gradually pour enough water into the funnel to cover the compost sample. Record the amount of water added.
6. Stir gently, then let sit until the sample is saturated.
7. After the compost is saturated, release the clamp to allow excess water to flow into the graduated cylinder.
8. After the dripping stops, record the amount of water that is in the graduated cylinder.

ANALYSIS

1. Calculate how much water was retained in the 100-ml sample of compost or soil:

   __ml water retained/100-ml sample =

   water added (ml) – water drained (ml)
   from Step 5   *from Step 8*

2. Water holding capacity is expressed as the amount of water retained per liter of soil, so the next step is to multiply by 10 to convert from the 100-ml sample to a full liter:

water holding capacity (ml/l) = 10 x (__ml water retained/100 ml sample)
from equation above

3. Compare the water holding capacities of various types of soil, with and without compost added.

ORGANIC MATTER CONTENT

USE: To determine the organic and mineral fractions of a compost or soil sample.

BACKGROUND

When an oven-dry sample of soil or compost is heated to 500°C, organic matter is volatilized. These "volatile solids" make up the organic fraction of soil, including living biomass, decomposing plant and animal residues, and humus, the relatively stable end product of organic decomposition. The residue left after combustion is ash, composed of minerals such as calcium, magnesium, phosphorus, and potassium. In general, 50–80% of the dry weight of a compost represents organic matter that is lost during combustion.

Organic matter makes up a much lower percentage of the dry weight of soils. Most are less than 6% organic matter, with higher percentages occurring in bog soils. Surface soils have higher organic matter contents than subsoils because humus is formed through decomposition of accumulated residues of crops or natural vegetation. The most productive soils are rich in organic matter, which enhances their capacity to hold both water and nutrients in the root zone where they are available to plants.

MATERIALS
- 10-g sample of compost or soil
- porcelain crucible
- tongs
- desiccator (optional)
- laboratory oven, Bunsen burner, or hot plate

If a Bunsen burner or hot plate is used for combustion:

- goggles
- glass stirring rod
- fan or other source of ventilation

PROCEDURE

1. Weigh the porcelain crucible, then add about 10 g of compost or soil.
2. Dry the sample for 24 hours in a 105°C oven.
3. Cool in a desiccator (or a nonhumid location), and reweigh.
4. Ignite the sample by placing it in a 500°C oven overnight. Using tongs, remove the crucible from the oven, and again place it in a desiccator or nonhumid location for cooling. Weigh the ash. A pottery kiln can be used if a laboratory oven is not available.

    Another option is to ignite the sample using a Bunsen burner or a hot plate. To avoid breathing the fumes, set up a fan or some other type of ventilation system. Wearing goggles, heat the sample gently for a few minutes, then gradually increase the heat until the crucible turns red. Stir the compost occasionally, and continue the combustion until the sample becomes light colored and you can no longer see vapors rising.

ANALYSIS

Calculate the percentage of organic matter using the following equation:

$$\text{organic matter (\%)} = \frac{W_d - W_a}{W_d} \times 100$$

in which:
W_d = *dry weight of compost*
W_a = *weight of ash after combustion*

How does the organic matter content of your compost compare with that of the soils you tested? Does the organic matter content diminish during the composting process, or does it just change in form and chemical composition?

If you divide the percentage of organic matter by 1.8 (a number derived through laboratory measurements), you can get an estimate of the percentage of carbon in your sample:

$$\text{\% carbon} = \frac{\text{\% organic matter}}{1.8}$$

This may be useful if you know the C:N ratio and you want to figure out the percentage of nitrogen:

$$\text{\% nitrogen} = \frac{\text{\% carbon}}{\text{C:N}}$$

BUFFERING CAPACITY

USE: To determine whether adding compost to soil increases the soil's capacity to resist pH change.

BACKGROUND

Finished compost usually has a pH around neutral, in the range of 6–8. It also tends to have a high buffering capacity, meaning that it resists change in pH. Soils with high buffering capacities do not experience drastic pH fluctuations that may be detrimental to microbial life and plant growth. Buffering capacity needs to be taken into account when determining the amounts of lime, sulfur, or other chemicals that are applied to soil to alter its pH.

The buffering capacity of soil may be provided by either mineral or organic components. Quartz sand has almost no buffering capacity, so even small additions of acid will drop the pH of the sand and its drainage water. In contrast, a sand made of crushed limestone is highly buffered because it contains calcium and magnesium carbonates. The addition of organic matter such as compost tends to increase a soil's buffering capacity.

This procedure provides a way of demonstrating the concept of a buffer. Students may be surprised to discover that compost with pH near 7 can neutralize an acidic solution. They might think that the compost would need to be basic to counteract the acidity of the solution, or they might expect the pH of the compost to drop corresponding to the increase in solution pH.

MATERIALS
- funnel
- ring stand with attachment to hold funnel
- circular filter paper or coffee filter large enough to line funnel
- 125-ml samples of compost and sand
- 250-ml beaker
- acid solution: add 1 ml 1M H_2SO_4 to 500 ml distilled water
- pH meter, test kit, or litmus paper

PROCEDURE
1. Attach the funnel to the ring stand, suspended above one of the beakers.
2. Line the funnel with filter paper or coffee filter.
3. Measure the pH of the compost (see p. 54).
4. Measure the pH of the H_2SO_4 solution.
5. Place a 125-ml sample of compost into the funnel.
6. Slowly pour a few milliliters of H_2SO_4 solution into the compost in the funnel. Continue adding small amounts of acid, stopping as soon as the liquid begins to drain into the beaker below.
7. Test the pH of the drainage solution and of the compost sample.
8. Optional: Repeat Steps 4–7, using sand in place of compost, then using a 50/50 mixture of sand and compost.

ANALYSIS

Did the pH of the compost change after application of the H_2SO_4 solution? What about the sand? What happened to the pH of the acid as it filtered through the compost or sand?

What can you conclude about the buffering capacity of the compost and the sand? Which would be better capable of withstanding the effects of acid rain? Which would be more resistant to pH change through a soil amendment such as lime?

Research possibility: Finished compost is near neutral pH. Can you design an experiment to answer one or more of the following questions: Is compost detrimental to use on acid-loving plants such as blueberries or azaleas? Does compost buffer the soil pH, making it harder to provide acidic conditions? How does it compare to peat moss in this regard?

---

[1] The Respiration Test is adapted from Bartha, R., and D. Pramer. 1965. Features of a flask and methods of measuring the persistence and biological effects of pesticides in soil. *Soil Science* 100:68–70.

[2] The Phytotoxicity Bioassay is adapted from: Zucconi, F., A. Pera, M. Forte, and M. de Betoldi. 1981. Evaluating toxicity of immature compost. *BioCycle* 22(2):54–57.

6
COMPOST AND PLANT GROWTH EXPERIMENTS

Up to this point, we have concentrated primarily on the processes involved in converting organic wastes to compost. But, in addition to being an environmentally sound means of reducing wastes, composting has important applications in agriculture and gardening. For example, compost can be used as a soil amendment to enhance the physical characteristics and productivity of soil. It can also be used as a mulch around shrubs, trees, and other plants to reduce soil erosion, evaporation, and weed growth.

A compost that is mature and relatively free of contaminants, and has favorable physical and chemical properties, should enhance the growth of plants. But, there are many questions about the effect of specific composts on plants. Do different species respond differently to compost? Does the response of any particular plant species to compost depend on the type of compost being used? What is the ideal mix of compost and soil for growing plants?

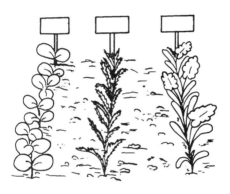

Plant growth experiments can provide answers to these and other questions students may devise. The experiments can be performed indoors in pots or outdoors in gardens and other field situations. Students also may develop research projects that combine plant growth experiments with some of the techniques presented in Chapter 5. For example, they may investigate the relationship of compost stability (as determined by the **Respiration Test**, p. 75) to the rate of growth of a particular species of plant. Or, they may design a research project to answer questions about how the porosity of a compost/soil mix relates to plant growth.

Research Possibility: Water in which compost has been soaked (often called compost tea) is said to be beneficial to plants. Can you design experiments to test whether different types, concentrations, and amounts of compost tea enhance plant growth?

The instructions below outline a protocol for conducting plant growth experiments in the laboratory. Students may want to adapt these protocols for use in greenhouses or outdoors. They can also use this or a similar protocol for any number of research projects, varying factors such as the type and maturity of compost, mixture of compost and soil, plant species, and environmental conditions (e.g., moisture, temperature, sunlight) under which the plants are grown. It is important to keep in mind that varying only one factor at a time makes a simpler student experiment, the results of which can be more readily interpreted. For example, if a student is interested in the effect of different compost/soil mixes on plant growth, it would be best to start with only one type of compost, one type of soil, and one species of plant, varying only the ratio of compost to soil. A later experiment or an experiment by another student could investigate another type of soil or compost, or a different species of plant.

Suppose a student varied both the ratio of compost to soil and the type of compost in the same experiment. It would be difficult to determine whether any differences in plant growth were due to the type of compost or to the relative amounts of compost and soil used. It is possible using statistical analyses to interpret experiments with more than one independent variable (e.g., ratio of compost to soil, type of compost). However, it is difficult for a beginning researcher to make sense of such results. Thus, we recommend that high school students stick to one independent variable at a time when first embarking on controlled experimental research.

Students should be careful in how they set up their experiments and how they interpret their results. For example, suppose a student wants to find out the effect of compost on plant growth. S/he conducts an experiment in which half the plants are grown in compost and the other half in soil. Reasoning that compost has a higher nutrient content than soil, s/he corrects for this by fertilizing only the plants grown in soil. If these plants grow faster than those grown in compost, what can s/he conclude? It would be reasonable to conclude that the soil/fertilizer combination provides a better growth medium than compost for this type of plant. But, what about the original question regarding the effect of compost on plant growth? Do plants grow better in a compost/soil mix than in plain soil? Did the plants in soil grow better because of the properties of the soil versus those of the compost, or because of the added fertilizer? A better experiment would be to grow plants in varying ratios of compost to soil, with fertilizer supplied either to all or none of the plants. Then, any differences in plant growth could be attributed to the various combinations of compost and soil in the potting mixes.

PLANT GROWTH EXPERIMENTS

USE: To determine the effect of compost on plant germination and growth.

MATERIALS
- pots or planting trays
- compost
- soil
- seeds
- light source (sunlight or artificial lighting)

PROCEDURE

1. Design your own experiment. There are many possibilities—a few ideas are listed here, but the variations are endless:
   - Test various combinations of soil and compost on plant growth. For example, you might want to dig a soil sample from your school yard and mix it with various amounts of finished compost for planting experiments. (Natural soil is better than prepackaged potting soils for experiments such as this because the potting mixes are formulated for optimal plant growth and already contain significant amounts of compost or humus.)
   - Another possibility is to mix your own potting soil by using vermiculite, sand, and compost. Creating several mixtures using the same percentages but different types of compost is a good way of comparing the influence of the various types of compost on plant growth. For example, you could compare compost at various levels of maturity, compost created using different mixtures of organic wastes, or vermicompost versus compost created in a thermophilic system.
   - If you are interested in investigating the effects of compost tea on plant growth, you could fill the pots with a sandy soil or potting medium such as vermiculite, then use compost extracts for watering.

2. Whatever type of experiment you choose, make sure that you design your experiment to include replicates of the various treatments. For example, your design might look like the following:

Treatment (% soil/% compost)	# Flats (with 6 plants in each)	# Plants
100% compost	3	18
25/75	3	18
50/50	3	18
75/25	3	18
100% soil	3	18

3. Plant your seeds, water them, and place them in a well-lit location. Many type of seeds will work, but radish or lettuce are often chosen because they grow quickly. Melon seeds are sensitive to fungal diseases,

and thus they provide a sensitive indicator of whether fungi have been killed through heating or curing of the compost.

4. Keep all the pots in the same setting to minimize any variation in temperature, lighting, pests, and other environmental factors. Even when the environmental conditions are kept as constant as possible, it is a good idea to randomize the grouping of plants rather than placing all the plants that are receiving the same treatment together in one group. This helps to further minimize the effect of any environmental differences.

5. Record on a daily basis the number of seeds that have germinated, plant growth, and observations about plant health such as color, vigor, or damage due to pests and diseases. You can decide what measurements to use as indicators of plant growth; possibilities include plant height, number and size of leaves, and dry weight of the entire plant at the end of the experiment. (For dry weight, weigh the plant after drying in a 105°C oven for 24 hours.)

ANALYSIS AND INTERPRETATION

1. Graph germination rates and plant growth over time for the different treatments. Also, determine the mean number of seeds germinated and mean size or mass of the plants at the end of the experiment. Compare average germination rates, plant growth, and health for the different experimental treatments. Based on your experiments, what was the optimal potting mix for plant germination? For plant growth? For plant health?

2. Some things may have gone wrong in your experiments. For example, you may have over-watered your plants, causing them all to die from fungal infection regardless of the treatment. Or you may have taken measurements only on plant height, and later decided that measuring the number of leaves and length of the main stem would have given better information. These types of problems are normal and can be used as a basis for redesigning the experiment. How might you change your experimental design if you were to carry out another set of growth experiments?

3. You may not find any differences between the treatments. Or, you may discover that the plants grown without compost did best. If this is the case, it may be difficult to determine whether the compost had no effect, or you did something wrong. The tendency is to assume the compost really has an effect and to attribute insignificant or negative results to experimental mistakes. However, the interpretation of results should not be biased by your predictions or preconceived ideas about the way experiments will turn out. Often unexpected results lead to important insights and questions. Maybe your compost is of poor quality, or maybe the plant species you chose grows well in poor soils. Explore all the possibilities for explaining your results with an open mind, through discussions and new experiments.

4. The conclusions and recommendations that you are able to make based on your results will depend on how and where you carried out your experiments. For example, if you used potted plants in a class-

room or greenhouse, it may be difficult to extrapolate from your results to what would happen if the same plants were grown outdoors in a garden. However, your results may give you some ideas about what would happen, allowing you to make predictions or hypotheses. You could then use these predictions to design a new experiment on plant growth in a garden setting.

7
COMPOSTING RESEARCH

Vermiculturists assert that worm castings are the most valuable commodity in the garden and that you can take raw pig manure and run it through worms to totally remove the pathogens and add countless new minerals while improving texture. Non-worm composters insist that worms lock up nutrients, destroy microorganisms, and result in a product that is less nutritive... My bottom line is simple, I am trying to determine what has been proven in order to arrive at an approach that integrates the best of what we know—without hype or speculation...
—E-mail message from frustrated compost manager

Welcome to the frontiers of science. I know this can be frustrating, but the debate and contradictions you experience are not likely to be resolved quickly... I recommend you approach the field of composting with skepticism as well as an open mind. An open mind is essential to understand some of the great composting mysteries that are almost certainly out there, perhaps related to the role of microbes or various unconventional practices. But skepticism is also critical, especially when listening to those who are certain they already know all the answers.
Best wishes on your journey!
—Response from compost scientist Tom Richard

Throughout *Composting in the Classroom*, we not only have presented information but also have pointed out gaps in our knowledge. And, even the information we have presented represents only the most current understanding about composting. As new information is added, our understanding of composting processes will be enhanced, and recommendations for composting procedures will change.

Maybe your students are intrigued by one of the research possibilities we have presented. Or, maybe they have questions of their own that could be developed into a research project. Perhaps their past experiences with composting caused them to doubt some of the information that has been presented, and they would like to test the validity of their own observations.

This final chapter is for those students who have posed questions in the course of exploring material in this manual and are eager to conduct research to find answers. Many of the techniques presented in Chapters 3–5, as well as the background information presented in Chapter 1, should be useful in designing and conducting experiments. The discussion of plant growth experiments in Chapter 6 has already provided an introduction to some of the issues encountered in conducting research.

This chapter presents a short overview of research as conducted by scientists, as well as some example research projects. Students can use the example research possibilities presented throughout this publication as models for their research, or they may devise entirely new projects.

EXPLORATION AND CONTROLLED EXPERIMENTS

Students and scientists working in a new field often start off with *exploratory research* (Figure 7–1). In this type of research, students or scientists are *exploring* or gaining a "feel for" an organism or process by using a variety of methods. An example of exploratory research would be students trying various mixtures of compost ingredients, testing their effects on the peak temperature reached and the length of time the temperature remained elevated. Another example of exploratory research would be students comparing the microbial populations in different composts. Or, students interested in the physical sciences could investigate how different bioreactor designs affect the temperatures achieved during the composting process. Such explorations would lead to a series of observations about the variables that could be changed or investigated in a more controlled manner in order to answer a more defined research question.

Based on results obtained during their exploratory research, students might refine their research questions and plans, leading to design and execution of a controlled scientific experiment. *Controlled experiments*[1] are carefully designed to include clearly defined objectives and hypotheses, dependent and independent variables, one or more treatments or levels of the independent variable, and replicates for each treatment. Based on their exploratory research, students studying compost microbiology might hypothesize that composts with high levels of nitrogen support more microbial activity. The next step would be to design an experiment to test this hypothesis, by purposefully varying the nitrogen content of the compost mix while keeping all other variables constant. Similarly, the students investigating bioreactors might have discovered through their exploratory research that systems smaller than a certain size lose heat too quickly to achieve the optimal temperatures for composting. These students could design a controlled experiment focusing on whether insulation can compensate for the larger surface-to-volume ratio in small compost systems. Or, they might choose to investigate whether the addition of an outside energy source enhances thermophilic composting in small bioreactors. In either case, the students would vary only one independent variable (insulation or outside energy source), keeping the size of the pile, mix of ingredients, ambient temperature, and other factors constant.

It should be noted that not all exploratory research leads to controlled experiments. Some students may not be ready to move on to controlled experiments, and they may be better served by conducting several exploratory projects. Because it is so open-ended and allows students so much freedom in designing their projects, exploratory research can serve to motivate students with differing abilities. Classroom discussions of the results of students' research projects can guide the teacher in determining when to introduce more advanced concepts such as replication and controls, and when to allow students to conduct further exploratory research in a more open-ended manner.

Figure 7–1. Model for Conducting Research.

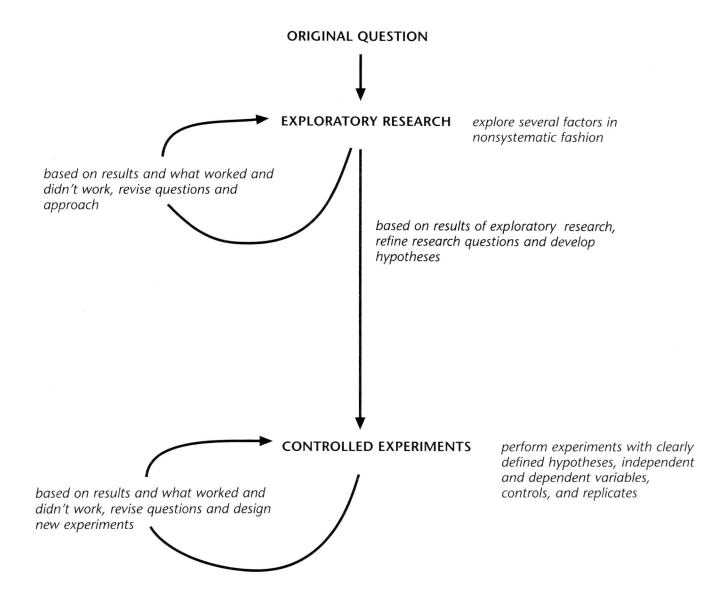

NARROWING DOWN A RESEARCH QUESTION

One of the most difficult aspects of conducting research is framing a research question that can be answered using the means available. In exploratory research, this is less important, because the goals are for students to become acquainted with a particular process or organism and for them to get excited about research. The results of exploratory research can often be used to identify trends, but they are not expected to yield solid conclusions. In fact, exploratory research is often used to help narrow down a question that can be answered by a controlled experiment.

The goal of controlled experimental research is to answer specific questions. However, the results of a controlled experiment will be difficult to interpret if the original question is not well defined.

For example, suppose that your students wanted to know whether compost enhances the growth of plants. They might start off by bringing in samples of home compost and using it as mulch around their favorite plants in the school yard. Once their results are in, it may be difficult to conclude why some plants grew faster than others. Was it because of differences in the species of the plant, or whether they were growing in the sun or shade? Was plant growth affected by the type of compost applied, or the amount? Although it will be hard to draw conclusions with so many variables, the students may notice a particularly interesting result or trend that stands out. For example, all the plants near the south-facing wall of the building grew particularly well regardless of the compost treatment. Could this be due to higher temperatures in that area of the school yard? You could use the students' observations to help them design a new, controlled experiment on the effect of temperature on plant growth.

Alternatively, you could go back to the original question about the effect of compost on plant growth, and use the results of the experiment to help the students narrow down a better research question. An example of a better-defined research question is, "What is the effect of vermicompost produced from cafeteria wastes on the growth of kidney beans?" Students can further narrow this question to: "What is the effect of adding different amounts of vermicompost to school-yard soil on the growth of kidney beans in pots?" They have now defined a question that can be answered in a controlled experiment, by varying the ratio of vermicompost to soil and growing the kidney beans under the same watering, light, and temperature regime.

EXAMPLE RESEARCH PROJECTS

Some ideas for research projects have been presented throughout this manual. We encourage students to use these and their own ideas to develop explorations and experiments. The following is an example of how the same question might be examined through both exploratory research and controlled experiments.

How long does it take for different composts to become stable?

Exploratory Research: Students could mix together various organic materials and measure the compost respiration rates at several-day intervals after the thermophilic stage has been completed.

Controlled Experimental Research: Students could design an experiment with carefully measured quantities of various ingredients. They would choose only two ingredients and vary their relative amounts. The total volume of wastes, ambient temperature, and moisture content should be kept constant. There would be at least three replicates of each treatment. From each replicate, respiration would be measured at each sampling time. The results would be presented as the mean (and standard error if students are familiar with this measure) of the measurements for each treatment at each sampling time. Any unexpected results should be presented.

The following are two examples of how an exploratory research project can lead into a controlled experiment.

How does the compost/soil mix affect the growth of lettuce seeds?

Exploratory Research: Students could start off by growing lettuce seeds in a variety of compost/soil mixes and measuring various properties of the different mixes (e.g., porosity, water holding capacity, pH). They might observe that lettuce seeds grow best within a narrow range of pH, regardless of the porosity or water holding capacity.

Controlled Experimental Research: Students could define a new question based on the results of their exploratory research. What is the effect of the pH of a compost/soil mix on the growth of lettuce seeds? They could then systematically vary the pH of a single compost/soil mix by using different amounts of lime or acid, and measure the growth of lettuce seeds.

What is the effect of physical factors (e.g., air flow) on peak temperatures achieved in soda bottle composting?

Exploratory Research: Students could build soda bottle bioreactors with various passive and active aeration systems, moisture conditions, and insulation, and then measure the temperatures during composting. Upon comparing their results, they might conclude that the hottest temperatures were achieved by using 5–cm thick foam insulation and passive aeration.

Controlled Experimental Research: Students might recognize that because they varied both the insulation and aeration system at the same time, they could not determine from their exploratory research which factor was more important in promoting high temperatures. They could design a new experiment to compare the effect on temperature of several different thicknesses of insulation, keeping aeration and all other variables constant. Alternatively, they could compare the effect on temperature of several different aeration systems, keeping insulation and all other variables constant.

Often, research projects can be readily converted to technological design projects. In the example discussed above, students researched the effect on compost temperature of various soda bottle bioreactor designs. Another possibility would be to turn this into a technological design project by asking the students to design the most effective or hottest soda bottle reactor.

INTERPRETING RESULTS

Many people are under the impression that research always reveals definitive answers to the questions that are asked. In reality, many research projects only partially answer the original questions. Sometimes, as mentioned above, the original question was not defined well enough. Often, the researchers discover that their methods were inadequate and need to be refined. For example, in trying to determine the effect of moisture content on compost temperature, it might turn out that the range of moisture contents was too narrow to show any effect, and a broader range should be used in a subsequent experiment. Other times, something goes wrong with the experimental procedure. For example, a continuously recording temperature device may stop working in the night, or a custodian might move an experiment.

Even when everything goes according to plan, the results of an experiment may not turn out as expected. A researcher might ask whether adding sawdust or newspaper to vegetable scraps results in higher temperatures in two-can bioreactors. S/he might use five garbage cans with sawdust and five with newspaper. What if the results come out like this?

Trial #	Maximum Temperature (°C)	
	Compost with Sawdust	Compost with Newspaper
1	21.4	26.3
2	26.0	29.9
3	20.2	25.0
4	25.7	15.2
5	18.6	15.8
Mean	22.4	22.4

Can you draw any conclusions? At first, you might say that there are not any differences between newspaper and sawdust because the mean temperatures are the same. This may be true, but by stopping at this point, you may miss some important possibilities. If you examine your results more closely, you will notice that in the first three trials, the newspaper mix produced higher temperatures, and all of them ranged between 25°C and 30°C. But what happened with the last two trials? Was something different? Was the newspaper mostly glossy and colored instead of black and white? Was there a difference in the vegetable scrap mixture? Asking questions about what may have led to discrepant results is fundamental to conducting research. It can often lead to new questions, and in turn, new experiments. It can also help you to redesign an experiment to answer your original question. Carefully examining and using all your results, whether or not they support your original hypothesis, is an essential part of conducting research.

FINAL WORDS

In sum, research is both a cumulative and iterative process. Each experiment builds on what the researcher already knows from reading, talking with other scientists and carrying out previous investigations. And, many experiments lead to refining of the original questions and methods; an experiment may have to be repeated in slightly different form several times before a question is answered. Finally, the results of an experiment, even when seemingly ambiguous or contradictory, often lead to new insights, new questions, and new investigations.

Similar to research, teaching is a process of trying new things and building on past experiences. And, similar to composting, not all the answers are known about how best to engage students in inquiry-based science. We have presented what is currently known about composting science, as well as the experiences of teachers and scientists involved in developing composting research projects suitable for high school students. Continued research and classroom experience will help us to refine composting methods and to develop new ways for students to "learn science as science is practiced." You and your students can be part of this process.

Best wishes on your journey!

---

[1] For a thorough discussion of student research, including dependent and independent variables, treatments, and replicates, see Cothron, J. H., R. N. Giese, and R. J. Rezba. 1993. *Students and Research: Practical Strategies for Science Classrooms and Competitions*, 2nd ed. Kendall/Hunt, Dubuque, IA.

GLOSSARY

Acid – A substance with pH between 0 and 7.

Actinomycetes – A type of bacteria, distinguished by their branching filaments called mycelia. Include both mesophilic and thermophilic species. In composting, actinomycetes play an important role in degrading cellulose and lignin.

Adsorption – The attachment of a particle, ion, or molecule to the surface of a solid such as soil or compost.

Aeration – The process through which air in compost pores is replaced by atmospheric air, which generally is higher in oxygen.

Aeration, forced – Addition of air to compost using blowers, fans, or vacuum pumps.

Aeration, passive – Relying on natural forces, such as convection, diffusion, wind, and the tendency of warm air to rise, for movement of air through compost.

Aerobic – (1) Characterized by presence of oxygen, (2) Living or becoming active in the presence of oxygen, (3) Occurring only in the presence of oxygen.

Air-dry – (adj) The state of dryness of a compost or soil at equilibrium with the surrounding atmosphere. (v) To allow to reach equilibrium in moisture content with the surrounding atmosphere.

Amino acid – An organic compound containing amino (NH_2) and carboxyl (COOH) groups. They combine to form proteins.

Anaerobic – (1) Characterized by absence of oxygen, (2) Living or functioning in the absence of oxygen, (3) Occurring only in the absence of oxygen.

Annelid – A member of the phylum Annelida, containing segmented worms.

Bacteria – Single–celled microscopic organisms lacking an enclosed nucleus. Members of the kingdom Monera. Commonly have a spherical, rod, or spiral shape.

Base – A substance with pH between 7 and 14.

Batch composting – Composting in which all of the ingredients are added at once rather than continuously over a period of time.

Bedding – A moisture-retaining medium for worm composting, such as shredded paper, leaves, or peat moss.

Bioassay – A laboratory procedure using living organisms to test the toxicity of a substance.

Biodegradable – Capable of being broken down through biochemical processes.

Biofilter – A filter that uses microbial action to reduce odors. Finished compost commonly is used as a biofilter to reduce potential odors from active compost systems. This can be as simple as layering finished compost over a pile containing fresh food scraps. In systems using forced aeration, the air commonly is blown through a biofilter of finished compost before being released to the environment.

Biomass – The mass of living organisms.

Bioreactor – An enclosed container used for making compost or conducting scientific experiments on the composting process.

Buffer – A substance that resists rapid change in pH.

Buffering capacity – The ability to resist change in pH.

Bulking agent – A material used in composting to maintain air spaces between particles.

C:N – See Carbon-to-nitrogen ratio.

Carbon-to-nitrogen ratio – The ratio of the weight of organic carbon to the weight of total nitrogen in soil, compost, or other organic material.

Castings – Worm feces, including undigested organic matter, bacteria, and soil that have passed through worm digestive systems.

Cellulose – The chief component of plant cell walls, cellulose is a series of organic compounds containing carbon, hydrogen, and oxygen formed into chains of 1000-10,000 glucose molecules. Cellulose forms the fibrous and woody parts of plants and makes up over 50% of the total organic carbon in the biosphere.

Clay – (1) A soil component consisting of the smallest mineral particles, those <0.002 mm in diameter, (2) Soil composed of >40% clay, <45% sand, and < 40% silt.

Compost – (v) To decompose organic materials under controlled conditions, (n) The humus-like material produced by decomposing organic materials under controlled conditions.

Compost maturity – See Maturity.

Compost quality – The suitability of a compost for use with plants. Compost that impairs seed germination or plant growth is of low quality, either because it is not yet fully decomposed or because the initial ingredients contained contaminants that are phytotoxic.

Compost stability – See Stability.

Compost system – The method used to convert organic wastes into a stable end product. Examples range from large outdoor windrows or piles to small indoor bioreactors.

Compost tea – An extract made by soaking finished compost in water.

Compound microscope – A microscope with two sets of lenses, one in the eyepiece and the other in the objective.

Conduction – The transfer of heat by physical contact between two or more objects or substances.

Control – In a scientific experiment, the test group used as a standard of comparison, to which no experimental factors or treatments have been imposed.

Convection – The transfer of heat through a gas or solution because of molecular movement.

Culture – (v) To grow organisms under controlled conditions, (n) The product of growing organisms under controlled conditions, such as a bacterial culture.

Curing – The final stage of composting, after the period of rapid decomposition has been completed, in which slow chemical changes occur that make the compost more suitable for use with plants.

Cyst – A sac surrounding an animal or microorganism in a dormant state.

Debris – Dead organic matter.

Decomposer – An organism that feeds on dead organic matter and aids in its degradation.

Dependent variable – In a scientific experiment, the variable that changes as a result of treatment variations. For example, in plant growth experiments, one dependent variable might be the height of the plants grown in potting mixes containing various amounts of compost.

Detritus – Dead organic matter.

Endospore – A structure inside some bacterial cells that is highly resistant to heat and chemical stress and can germinate to grow a new cell when environmental conditions become favorable.

Enzyme – An organic substance produced by living cells and capable of acting as a catalyst for a biochemical reaction.

Feces – Wastes discharged from the intestines of animals.

Food chain – A hierarchical sequence of organisms that feed on each other, starting with either green plants or organic detritus as the primary energy source.

Food web – The network of interconnected food chains found in an ecological community.

Forced aeration – See Aeration, forced.

Fungi – Plural of fungus. A kingdom that includes molds, mildews, yeasts, and mushrooms. Unlike bacteria, fungal cells do have nuclei. Fungi lack chlorophyll, and most feed on dead organic matter. In compost, fungi are important because they break down tough debris like cellulose, and they grow well during the curing stage, when moisture and nitrogen levels are low.

Gram negative – A characteristic of bacterial cells that are decolorized by 95% alcohol during the Gram staining procedure.

Gram positive – A characteristic of bacterial cells that retain Gram stain color after washing with 95% alcohol during the Gram staining procedure.

Gram staining – A differential staining procedure that provides for separation of bacteria into gram-positive and gram-negative types, depending on the structure of their cell walls.

Heavy metals – Metallic elements with high molecular weights. Includes cadmium, lead, copper, mercury, chromium, silver, and zinc. High concentrations in soil can be toxic to plants or to animals that eat the plants or soil particles.

Hemicellulose – A series of organic compounds made up of chains of 50–150 sugar units including glucose, xylose, and galactose. In wood, hemicellulose surrounds cellulose and helps to bind it to lignin.

Holding unit – A simple container that holds landscaping, garden, and food wastes while they break down.

Humus – The stable organic complex that remains after plant and animal residues have decomposed in soil or compost.

Hydrolysis – A chemical reaction in which water is one of the reactants.

Hyphae – Branched or unbranched chains of cells, as in fungi and actinomycetes.

Hypothesis – A prediction about the relationship between variables in a scientific experiment.

Immature compost – See Maturity.

Independent variable – A factor that is intentionally manipulated in a scientific experiment. For example, in plant growth experiments, one independent variable might be the ratio of compost to sand in the planting mixture.

Inoculant – Microorganisms that are introduced into compost or other culture media.

Inoculate – To introduce pure or mixed cultures of microorganisms into culture media.

Inorganic – Mineral, rock, metal, or other material containing no carbon-to-carbon bonds. Not subject to biological decomposition.

Invertebrate – An animal without a backbone, such as an insect or worm.

Leachate – The liquid extract that results when water comes into contact with a solid such as soil or compost. In composting, leachate containing dissolved and suspended substances drains from the system as organic matter decomposes.

Lignin – A series of complex organic polymers that are highly resistant to microbial decomposition. In wood, lignin cements cellulose fibers together and protects them from chemical and microbial decomposition.

Lime – Calcium compounds used to neutralize acidity in soils.

Macrofauna – Soil-dwelling invertebrates that are large enough to create their own burrows.

Macroorganism – An organism large enough to be observed with the naked eye.

Mature compost – See Maturity.

Maturity – A measure of whether compost has completed not only the phase of rapid decomposition, but also the longer curing phase during which slow chemical changes make the compost more suitable for use with plants.

Mesofauna – Soil-dwelling invertebrates that are intermediate in size. They live in the air-filled pores between soil or compost particles but generally do not create their own spaces by burrowing.

GLOSSARY

Mesophilic – (1) Organisms that grow best at moderate temperatures (10–40°C), (2) The phase of composting that takes place at temperatures in the range of 10–40°C, (3) The type of composting that does not reach temperatures exceeding 40°C.

Microfauna – Soil protozoa and other microscopic fauna that are small enough to live in the thin film of water surrounding soil or compost particles.

Microbe – A microorganism.

Microorganism – An organism that individually is too small to be observed without magnification through a microscope.

Mulch – Any material such as compost, bark, wood chips, or straw that is spread on the soil surface to conserve soil moisture, suppress weed growth, moderate temperature changes, or prevent soil erosion.

Mycelia – Branching networks of fungal hyphae.

Nitrifying bacteria – Bacteria that transform ammonium (NH_4^+) to nitrate (NO_2^-) and then to nitrate (NO_3^-).

Nitrogen-fixing bacteria – Bacteria that transform atmospheric nitrogen (N_2) to ammonium (NH_4^+).

Organic – (1) Pertaining to or derived from living organisms, (2) Chemical compounds containing carbon-to-carbon bonds.

Organic matter – Material that has come from something that is or was once alive.

Oxidation – A chemical reaction in which an atom loses electrons or increases in oxidation number.

Passive aeration – See Aeration, passive.

Pathogen – Any organism capable of producing disease or infection in other organisms.

Percolation – Downward movement of water through pores in rock, soil, or compost.

Permeability – The ability of a soil to allow the movement of water through its pores.

pH – The degree of acidity or alkalinity of a substance, expressed as the negative logarithm of the hydrogen ion concentration. Expressed on a scale from 0 to 14. pH <7 is acidic, 7 is neutral, and >7 is alkaline or basic.

Phytotoxicity – A measure of the ability of a substance to suppress seed germination, injure plant roots, or stunt plant growth.

Polymer – A large, chain-like molecule composed of many identical repeating units.

Pore – An open area between particles of compost or soil, filled by air or water.

Porosity – The percentage of the total soil or compost volume that is occupied by open spaces rather than solid particles.

Protozoa – Single-celled, animal-like microorganisms belonging to the kingdom Protista. Many species live in water or aquatic films surrounding soil or compost particles.

Radicle – An embryo root that grows when a seed germinates.

Replicate – In a scientific experiment, the experimental units to which the same treatment has been imposed. For example, in a planting experiment, five plants would be replicates if they were set up under the same initial conditions and received the same treatment throughout the experiment.

Sand – (1) A soil component consisting of mineral particles between 0.05 mm and 2.0 mm in diameter, (2) Soil composed of >85% sand and < 10% clay.

Screening – The process of passing compost through a screen or sieve to remove large pieces and improve the consistency and quality of the end product.

Silt – (1) A soil component consisting of mineral particles between 0.002 mm and 0.05 mm in diameter, (2) Soil composed of >80% silt and < 12% clay.

Soil amendment – Any substance that is used to alter the chemical or physical properties of a soil, generally to make it more productive. Examples include compost, lime, sulfur, gypsum, and synthetic conditioners. Usually does not include chemical fertilizers.

Specific heat – The quantity of heat needed to raise the temperature of 1 g of a substance by 1°C.

Stability – A measure of whether compost has decomposed to the point at which it does not reheat, produce offensive odors, or support high rates of microbial growth when optimal moisture levels are supplied.

Temperature profile – A graph of temperature changes that occur during the process of thermophilic composting.

Thermophilic – (1) Organisms that grow best at temperatures above 40°C, (2) The phase of composting that takes place at temperatures exceeding 40°C, (3) The type of composting that includes a stage occurring above 40°C.

Turning – In a compost pile, mixing and agitating the organic material.

Turning unit – Three holding units built next to each other, so that compost can be turned from one into the next.

Variable – A factor that changes in a scientific experiment.

Vermicompost – (v) To decompose organic matter using worms, (n) The product obtained through decomposition of organic matter by microorganisms and worms.

Vermiculite – A mineral material that is used in potting soil to keep the mixture light and porous.

Windrow – An elongated pile of organic matter in the process of being composted.

Yard and garden wastes – Grass clippings, dead leaves, small branches, weeds, and plant residues.

FOR MORE INFORMATION

COMPOSTING AS A TOPIC FOR HIGH SCHOOL SCIENCE INVESTIGATION

Trautmann, N.M., T. Richard, and M.E. Krasny. 1996. *Composting in Schools*. WWW site at:
http://www.cfe.cornell.edu/compost/schools.html

This WWW site includes detailed information on compost science and engineering, as well as articles about weird and unusual composting, frequently asked questions, a composting quiz, and bulletin boards for posting messages to other teachers or students.

Krasny, M.E. and N.M. Trautmann, executive producers. 1996. *It's Gotten Rotten*. Produced by Photosynthesis Productions, Inc. Ithaca, NY.

A 20–minute video designed to introduce high school students to the science of composting, highlighting the biology of the invertebrates and microorganisms that decompose organic matter. Students are shown designing and using both indoor and outdoor composting systems, observing living organisms, and using finished compost to grow plants. Available from:

Cornell University Resource Center or Bullfrog Films
7 Business & Technology Park   Box 149
Ithaca, NY 14850   Oley, PA 19547
Phone (607) 255–2090   Phone (800) 543–FROG
Fax (607) 255–9946   Fax (610) 370–1978
E-mail: Dist_Center@cce.cornell.edu E-mail: bullfrog@igc.org

COMPOSTING FOR WASTE MANAGEMENT

Bonhotal, J.F. and K. Rollo. 1996. *Composting...Because a Rind Is a Terrible Thing to Waste*. Cornell Waste Management Institute. Ithaca, NY.

A handbook and two videos designed to help institutions such as schools to implement food scrap composting. Available from Cornell University Resource Center (address above).

Bonhotal, J.F. and M.E. Krasny. 1990. *Composting: Wastes to Resources*. Cornell Cooperative Extension. Ithaca, NY.

A guide for educators and volunteers leading youth aged 9–12 in composting and related activities. Includes plans for eleven types of outdoor composting systems. Available from Cornell University Resource Center (address above).

Dickson, N., T. Richard, and R. Kozlowski. 1991. *Composting to Reduce the Waste Stream: A Guide to Small Scale Food and Yard Waste Composting*.

Explains how to construct and maintain a compost pile. Outlines factors that affect the composting process including aeration, moisture, and temperature. Illustrations, tables, diagrams, and step-by-step instructions for constructing compost bins. Available from:

NRAES
152 Riley-Robb Hall
Cornell University
Ithaca NY 14853
Phone (607) 255–7654
Fax (607) 254–8770
E-mail: nraes@cornell.edu

Martin, D.L. and G. Gershuny, eds. 1992. ***The Rodale Book of Composting***. Rodale Press. Emmaus, PA.

An updated version of Rodale's classic book for small-scale outdoor composting. Includes composting instructions, as well as the history of composting, relationships between compost and plant health, and recommendations for applying compost to lawns, gardens, and house plants.

Witten, M. 1995. ***Scraps to Soil: A How-To Guide for School Composting.*** Association of Vermont Recyclers. Montpelier, VT.

A guide for teachers and students in grades 3–6 who are interested in composting their school's food and landscaping wastes. Although not aimed at the high school level, the guide outlines the process for setting up a school-wide food scrap composting program. Available from:

Association of Vermont Recyclers
PO Box 1244
Montpelier, VT 05601
Phone (802) 229-1833

CHEMISTRY

Golueke, C.G. 1992. Bacteriology of composting. ***BioCycle*** 33(1): 55–57.

Kayhanian, M. and G. Tchobanoglous. 1992. Computation of C/N ratios for various organic fractions. ***BioCycle*** 33(5): 58–60.

PHYSICS

Haug, R. 1993. Thermodynamic Fundamentals. pp. 95–120 in ***The Practical Handbook of Compost Engineering***. Lewis Publishers, Boca Raton, FL.

MICROBIOLOGY

Alexander, M. 1991. ***Introduction to Soil Microbiology***, 2nd ed. Krieger Publishing Co., Malabar, FL.

Beffa, T., M. Blanc, and M. Aragno. 1996. Obligately and facultatively autotrophic, sulfur- and hydrogen-oxidizing thermophilic bacteria isolated from hot composts. ***Archives of Microbiology*** 165: 34–40.

Golueke, C.G. 1992. Bacteriology of Composting. ***BioCycle*** 33(1): 55–57.

INVERTEBRATES

Coleman, D.C. and D.A. Crossley, Jr. 1996. ***Fundamentals of Soil Ecology***. Academic Press, San Diego.

Dindal, D.L. 1990. ***Soil Biology Guide***. John Wiley & Sons, New York.

Dindal, D.L. 1971. ***Ecology of Compost: A Public Involvement Project***. State University of New York, College of Environmental Science and Forestry. Syracuse, NY.

The classic reference on food webs in compost piles. Available from:

Office of News and Publications
122 Bray Hall
ESF
1 Forestry Drive
Syracuse, NY 13210
Phone (315) 470–6500

Johnson, C.E. 1980. The wild world of compost. ***National Geographic*** 157: 273–284.

Mallow, D. 1990. Soil arthropods. ***The Science Teacher***. 57(5): 64–65.

WORMS

Appelhof, M. 1982. ***Worms Eat My Garbage***. Flower Press, MI.

The original book of instructions for setting up and maintaining a worm composting system. Available from:

Flower Press
10332 Shaver Rd.
Kalamazoo, MI 49002
Phone (616) 327–0108

Edwards, C.A. and P.J. Bohlen. 1996. ***Biology and Ecology of Earthworms***. Chapman Hall. London, UK.

STUDENT RESEARCH

Cothron, J.H., R.N. Giese, and R.J. Rezba. 1993. ***Students and Research: Practical Strategies for Science Classrooms and Competitions***, 2nd ed. Kendall/Hunt, Dubuque, IA.

CURRENT COMPOSTING RESEARCH

Compost Science & Utilization. The JG Press, Inc. Emmaus, PA.

A quarterly peer-reviewed journal that presents results of research on the science of compost production, management, and use.

INDEX

A

Actinomycetes 13, 16, 60–61
Aeration 10–11
Algebra 47–50
Ammonia 6, 8, 53
Anaerobic decomposition 8, 10, 52–53, 73
Annelids 19, 22–26
Ants 20
Arachnids 19
Arthropods 19–20
Aspergillus fumigatus X

B

Bacteria 13, 14–16, 57–61
Bedding for worms 36
Beetles 13, 20
Bins, see:
    Enclosed bins
    Holding units
    Turning units
    Worm bins
Biology 13–26
Bioreactors 3, 27–40
Buffering capacity 71, 72, 83, 89–90
Bulking agents 11, 31, 46

C

C:N ratio 6–7, 44–45, 48–50
    Calculations 48–50
    Of common compost ingredients 45
    Of tree leaves 45
Calculations for thermophilic composting 47–50
Carbon dioxide 5, 7, 10, 75–78
Cellulose 3, 5–7, 16, 21
Centipedes 13, 17, 20
Chemistry 5–8
Conduction 9
Convection 9
Crustaceans 20
Curing 2, 3, 16, 30, 34, 42, 64, 71

D

Data interpretation 91–92, 94–95, 102

E

Earthworms 17, 19, 22–26
Earwigs 17, 20
Eisenia fetida 22–26, 35–40
Enchytraeids, see Potworms
Enclosed bins 42
Enzymes 5, 15, 16, 21
Experimental design 91–93, 98–101

F

Fats 3, 6, 37
Flies 1, 20, 31, 38, 51
Food web in compost 13–14
Fruit flies 38
Fruit scraps X, 36–37, 43, 45
Fungi IX–X, 5, 8, 13, 15, 16–17, 62–63

G

Grass clippings 30, 31, 32, 43, 44, 45, 47–50

H

Heat, see Temperature
Heat loss 9–10
Hemicellulose 3, 6
Holding units 41–42
Hot composting, see Thermophilic composting
Humus 5, 85, 87, 93

I

Insects 19–20
Invertebrates 17–26, 67–69

J

Jar Test 71, 72, 73

L

Leaves X, 19, 30, 32, 41–42, 43, 45, 46, 47–50, 51, 53
Lignin 5–6, 7, 16, 23
Lumbricidae 25–26
Lumbricus terrestris 23, 25

M

Mesophilic composting X, 2, 27
Mesophilic microorganisms 3, 15, 16, 27
Microorganisms 14–17, 55–65, see also:
    Actinomycetes
    Bacteria
    Fungi
    Protozoa
Millipedes 13, 17, 20
Mites 13, 17, 19, 39
Mixing, see Turning
Moisture 10–11, 43–44, 47–48
    Calculations 47–48
    Measuring compost moisture 44
    Monitoring 52–53
    Of common compost ingredients 43
Mollusks 21

N

Nematodes 13, 17, 20–21
Newspaper 7, 16, 30, 32, 36, 45, 52–53, 102

INDEX

O
Odor 51, 53
Oligochaetes, see Earthworms
Organic matter content 71, 72, 83, 87–88
Outdoor composting 41–42
Oxygen 1, 3, 6, 7, 10–11, 73

P
Particle size 7, 11
Pathogens X, 3, 24, 46
pH 6, 8, 16, 24, 26, 53–54, 89–90
Physics 9–12
Phytotoxicity bioassay 71, 72, 79–82
Pillbugs, see Sowbugs
Plant growth experiments 91–96
Porosity 71, 72, 83–84
Potworms 17, 19, 39
Proteins 1, 3, 5, 6, 8
Protozoa 13, 17
Pseudoscorpions 13, 17, 19

Q
Quality (compost quality) 71–72

R
Radiation 10
Red worms, see Earthworms
Research VII–VIII, 91–96, 97–103
Respiration test 71, 72, 75–78

S
Sawdust 11, 30, 43, 45, 46, 51, 102
Self-heating test 71, 72, 74
Size of compost system 3, 12
Slugs 17, 21
Snails 17, 21
Soda bottle bioreactor 27, 31–34
Soil properties, see:
    Buffering capacity
    Organic matter content
    Porosity
    Water holding capacity
Sowbugs 13, 17, 20
Specific heat 10
Spiders 19
Springtails 13, 17, 19–20
Stability 71–72
Starches 5, 15
Sugars 5, 6, 15, 23

T
Temperature IX, 2–4, 9–10, 15, 29, 33, 51–52, 74
Temperature profile 2, 29, 33
Thermophilic composting IX–X, 2–4, 6–12, 15–16
Thermophilic microorganisms 15–16
Troubleshooting 38–39, 51

Turning 3–4, 10–11, 42, 53
Turning units 42
Two-can bioreactors 27–30

V
Vegetable scraps X, 36–37, 43, 45
Vermicomposting 22–26, 35–40

W
Water holding capacity 71, 72, 83, 85–86
Weeds 2, 46, 91
White worms, see Potworms
Wood chips 7, 11, 28–30, 32, 43, 45, 46, 51, 52, 53, 72
Worm bins 35–40
Worm composting, see Vermicomposting
Worms, see:
    Earthworms
    Potworms